Transducers for Microprocessor Systems

Transducers for Microprocessor Systems

J.C. Cluley

M.Sc., C.Eng., M.I.E.E., F.B.C.S.

Springer-Verlag New York Inc.

First published 1985 by
Higher and Further Education Division
MACMILLAN PUBLISHERS LTD
London and Basingstoke

Sole distributors in the USA and its Dependencies
Springer-Verlag New York Inc.
175 Fifth Avenue
New York, NY 10010
USA.

Printed in Hong Kong

ISBN 0-387-91268-1 Springer-Verlag New York

Contents

Preface

Many of the millions of microprocessors that are produced every year are used as the controlling elements in industrial plant, test equipment, domestic appliances, chemical plant, material handling equipment etc. To fulfil their allotted role the microprocessor systems need to collect data on external conditions and states such as temperature, pressure, force, velocity, torque and flow rate. These are not electrical quantities and thus transducers are required to transform these variables into electrical signals which the microprocessor system can deal with.

Also the microprocessor may be designed to modify some of these quantities in order to perform its control function; again transducers must be attached to the output circuits of the microprocessor.

The selection of suitable transducers is often a critical factor in the design of microprocessor systems, and a knowledge of the types available, their performance and relative cost is essential for engineers involved in design and development.

Unfortunately, since most transducers are not essentially electronic in nature, few courses in electronic engineering devote much time to a study of transducers and their applications. I have tried in this book to give an overall picture of the range and variety of transducers that may be needed in microprocessor systems, an outline of their main characteristics, and the way in which they can be interfaced to typical microprocessors.

For completeness I have included a chapter that gives a brief description of the more important interface packages used to connect transducers to the microprocessor. Readers requiring further details of interfacing or the way in which microprocessors handle information may find it helpful to consult companion Macmillan books: my own *Interfacing to Microprocessors*, and *Understanding Microprocessors* by B. S. Walker.

I hope that students of electronic engineering and practising engineers involved in the design of microprocessor systems and instrumentation will find in this book a useful introduction to the selection and application of transducers.

J. C. CLULEY

List of Abbreviations

ACIA	Asynchronous Communications Interface Adapter
ADC	Analogue-to-Digital Converter
BCD	Binary Coded Decimal
CMOS	Complementary Metal Oxide Silicon
CPU	Central Processor Unit
CR	Control Register
DAC	Digital-to-Analogue Converter
DDR	Data Direction Register
DMA	Direct Memory Access
LCD	Liquid Crystal Display
LED	Light Emitting Diode
LS	Least Significant
MS	Most Significant
PIA	Peripheral Interface Adapter
PIO	Parallel Input/Output Controller
PPI	Programmable Peripheral Interface
RAM	Random Access Memory (Writable Storage)
ROM	Read-Only Memory
VDU	Visual Display Unit

1

The Role of Transducers in Microprocessor Systems

1.1 Introduction

Many of the microprocessors now in service are used to control a wide range of plant, processes, machines or appliances. All of these applications involve close and continuous interaction between the microprocessor and its environment. Since the nature of the signals that the microprocessor can emit and receive is closely specified, information from the outside world generally requires a substantial degree of processing and conversion before it can be accepted by the processor. Equally any output from the processor usually needs modification before it is suitable for exercising control over plant, processes etc.

If we consider the input of information we can recognise the following stages in the process:

(a) Physical state or quantity.
(b) Electrical signal representing (a).
(c) Digitally encoded form of (b), usually represented by a parallel data word.
(d) Version of (c) with signal levels suitable for microprocessor input.
(e) The transmission of (d) into the microprocessor via the data bus, synchronised with the input instructions of the computer program.

In some cases one or two of these stages may not be required, depending upon the equipment used and the nature of the quantity being measured. Generally the transformation of the signal from each of these stages to that following requires some apparatus, as described below.

The generation of the electrical signal corresponding to the change from (a) to (b) is performed by a transducer or sensor. The function of these devices is described in more detail in section 1.2.

The change from (b) to (c) is performed by an analogue-to-digital converter (ADC), various types of which are described in chapter 2.

The change from (c) to (d) can often be avoided, but some details of where it is necessary are given in chapter 2.

The final stage from (d) to (e) is the function of the microprocessor interface, usually a special package designed to handle 16–20 lines to external devices. A brief account of the operations involved and the hardware used is given in chapter 3.

1.2 The function of transducers

The device that produces an electrical signal representing some physical state or quantity is usually called a *transducer*. Strictly speaking this term should be restricted to devices that change one form of energy into another, for example mechanical energy into electrical energy. Thus inevitably the device must abstract some energy from its environment in order to generate the electrical output.

However, it is often either essential or convenient to obtain the energy required for the output from an external supply, and not from the system or process being measured. Such devices should strictly be called *sensors*, although the terminology is not always used in its strict sense.

Some authors have used the term 'passive transducer' in place of sensor. For convenience the term transducer will be used in this book to include sensors also.

Many sensors use a beam of light to determine the position of an object or its movement, and so will not impose any load or disturbance upon it. Any energy that appears in the output signal can come only from the external power supply.

A third class of transducer is sometimes described as a 'feedback transducer'. This is really a description of the measuring circuits to which the transducer is connected, which are arranged as an error-driven negative feedback system. Measurement is usually performed by determining the magnitude of the signal that opposes the input quantity, rather than making a direct measurement.

For example, in a force balance system used for weighing, the pan is suspended in springs and any movement it makes is detected by an optical system. The error signal is amplified and connected to a moving coil motor which forces the pan upwards. In equilibrium the weight in the pan is balanced by the force produced by the moving coil motor, which is in turn proportional to the current flowing through it. The force is then measured by measuring the current in the moving coil, usually by a digital ammeter.

If it was possible, all transducers would be designed to produce digital output signals, so obviating the need for stage (b)–(c) of the signal transformation. Unfortunately the great majority of transducers are selected on other grounds such as low cost, reliability, accuracy etc., and produce an analogue output, preferably one in which the electrical output bears a linear relation to the physical state being measured.

Thus if we are measuring the angular velocity ω of a rotating shaft we would expect an output voltage of the form

$$K_1\omega + K_0$$

Here K_0 is the zero error, and K_1 is the scaling factor, expressed for example in volts per radian/s.

In many cases K_0 is zero, so simplifying the relationship between transducer input and electrical output, but this is not always possible. For example, if we wish to measure temperature, a convenient semiconductor device produces a current proportional to absolute temperature.

Thus to obtain a signal representing degrees Celsius it is necessary to offset the current corresponding to 0°C, so producing zero output at this temperature.

Although many transducers are linear, in some cases the physical relationships involved in the measurement produce an inherently non-linear characteristic. For instance, the velocity of a fluid may be measured by means of a transducer sensitive to the pressure drop across a venturi or constriction, which is proportional to the square of the velocity. Thus a linear pressure drop-to-voltage transducer will give a non-linear relation between fluid velocity and transducer output.

A key factor in any measurement is the interaction between the measuring instrument and its environment. In order to obtain an accurate value for any physical state or quantity, it is essential to make sure that the measuring device does not disturb the factor being measured. In particular, any abstraction of power or energy from the environment must be small enough not to affect it. Thus a tachometer generator that needs a mechanical input power of 10 watts could be attached to a 20 kW motor and would cause no significant increase in the load and consequently no material change in motor speed. However, were it coupled to a small motor delivering only 50 watts at full load, it would clearly produce a very significant increase in the load, and a marked change in speed.

1.3 Transducer dynamic response

The above example of coupling a tachometer to a small motor is one in which the power needed to drive the tachometer causes an error in the steady-state reading of the transducer. A similar error occurs if we use a low resistance voltmeter in a high impedance circuit. Here the error can easily be quantified. If the meter draws a current I amps, and the internal resistance of the circuit is R ohms, the measured potential will be $I \times R$ volts below that which would be measured on open circuit, by a meter that draws a negligible current.

Some transducers will produce no inherent error in a steady-state reading but because of their inertia will be unable to follow rapid fluctuations in the quantity being measured. This is the case when a conventional pressure transducer is used to measure the very rapid pressure changes that occur in the cylinder of a petrol engine.

In order to measure such fast changes accurately, special piezo-electric transducers were developed which could handle frequencies well into the kilohertz range, several orders of magnitude above those to which diaphragm and strain gauge transducers could respond.

Similar problems occur when trying to measure rapid changes of temperature. Unless thermocouples having a low thermal capacity are used, they will not be able to follow rapid changes in temperature and their outputs will always lag behind the changing temperature of their surroundings. Only when the temperature has remained constant for some time will the output be correct.

1.4 Transducer resolution

The signal conversion from an analogue or varying voltage to a digital value inevitably limits the smallest change that can be represented. Theoretically, an analogue voltage can be infinitely variable, whereas a digital value expressed by means of an N-bit binary number can represent only 2^N different (and normally equally spaced) values. Thus in a system with a range of 0 to V volts, the smallest voltage increment corresponding to a 1-bit change is

$$v = \frac{V}{2^N} \text{ volts}$$

In practice, all analogue voltages include a degree of uncertainty on account of transducer errors and non-linearity, as well as signal fluctuations caused by noise from amplifiers or induced in connecting leads. There is then little point in making the smallest voltage increment v markedly less than the combined uncertainty caused by the noise and transducer errors.

Equally there is no point in representing the information digitally with much more resolution than the application requires. If we consider for example the microprocessor control of a domestic washing machine, we may assume that the water temperature will be confined to the range 0–100°C, and that an uncertainty of say 2°C in the washing temperature would not materially affect the washing action. This means a resolution of 1 part in 50, so that a 6-bit digit giving a resolution of 1 part in 64 would suffice to represent the variable water temperature.

A 7-bit number would thus provide greater resolution than is strictly necessary. In practice, values may be represented by widely differing number sizes, depending upon the application. In the case of limit sensors where we need know only when a variable exceeds a specified value, only a single bit is required. At the other extreme a numerically controlled lathe may be required to turn a component 15 cm long to an accuracy of 0.0025 mm. This requires a resolution of 1 part in 60,000 which is just within the capacity of a 16-bit number.

Table 1.1 Minimum voltage increment and number of increments in conversion between digital and analogue signals

No. of bits	*No. of increments*	*Minimum voltage increment* Range 0–5 V	Range 0–10 V	Range ±10 V
6	64	78.1 mV	156 mV	313 mV
8	256	19.5 mV	39.1 mV	78.1 mV
10	1024	4.88 mV	9.77 mV	19.5 mV
12	4096	1.22 mV	2.44 mV	4.88 mV
14	16384	305 μV	610 μV	1.22 mV
16	65536	76.3 μV	153 μV	305 μV

Table 1.1 shows the smallest increment of voltage corresponding to a 1-bit change for a range of number sizes and full scale voltages. Since there is liable to be an error of at least ±1 bit when a signal is converted from analogue form to digital form, or in the reverse direction, this table gives an indication of the size of error that is unavoidable in such conversions.

Since the accuracy needed in the analogue circuits inevitably increases with the number of bits encoded, the cost of conversion increases rapidly with the bit size. Thus converters of moderate speed handling 8 bits cost somewhat less than most microprocessor packages, but increasing resolution to 12 or 14 bits increases the cost by an order of magnitude or more.

The implication is therefore that the system designer must always specify the minimum resolution that will meet the required performance criterion, to keep the overall cost down.

Although the conversion of analogue input signals to digital values (a necessary feature of all microprocessor systems) usually involves some loss of resolution, it can bring other advantages. Chief among these is the wide range of signal processing actions that the microprocessor can perform on the captured data. For example, we can take a number of readings of a particular variable in a short period of time during which it is expected to remain almost constant. By comparing each reading with the average value we can estimate the amount of random noise and disturbance that is superimposed on the signal. We can also use the average value to give a better estimate of the true value of the signal than can be obtained from any single reading.

1.5 Economic factors

The introduction of cheap microprocessor packages has enabled computing techniques to be applied to a very wide range of products, processes and appliances where the use of conventional small computers would be far too expensive.

However, there is no great economic advantage in reducing the cost of the digital processing by perhaps two orders of magnitude if expensive transducers are required, since the overall system costs still remain high.

The user of a microprocessor system will generally expect the cost of the external interface hardware and the transducers to be broadly comparable with the cost of the central processor and storage, and indeed if this is not the case many microprocessor applications are uneconomic.

There is thus a continuing demand for cheap transducers. Fortunately many applications do not call for the accuracy and resolution of laboratory equipment so that the combination of reduced standards of accuracy and much greater volume of production has enabled the price of many transducers to be reduced considerably. Such transducers can satisfy the requirements of comparatively inexpensive systems.

A further factor in the economic design of microprocessor systems for controlling products or plant is the cost of the apparatus into which they are fitted. Comparatively expensive transducers can generally be justified when microprocessor control is built into a plant costing, say, £50,000 since the benefits of improved efficiency or higher output will usually pay for the control in a year or two.

At the other extreme, the application of microprocessors to domestic products such as microwave ovens and washing machines is cost-effective only if inexpensive transducers can be used. Fortunately in many such applications most of the transducers required will have been part of the earlier electrical or electro-mechanical control systems. Thus their use with the microprocessor system does not involve any additional charge.

For example, washing machines require a timer and some means of measuring water temperature and sensing when the machine is full. A microprocessor controller does not generally offer any major improvement in the use of the machine, but it is increasingly being incorporated as it is much more reliable than an electro-mechanical timer and may be simpler to program. In this case no additional transducers are required for the microprocessor system.

The demand for inexpensive transducers has in some instances prevented the widespread use of microprocessor controls. For example, they can be used to control the carburettor of a petrol engine in a car to ensure an optimum fuel-to-air ratio for maximum engine efficiency and low pollution. However, this requires a cheap and reliable transducer for measuring the carbon monoxide content of the exhaust gases which will continue to perform correctly for perhaps 50,000 miles. As yet no such transducer has been developed and so mass-produced cars do not yet have microprocessor control of carburation.

This chapter has introduced the major topics associated with the use and selection of transducers. The next chapter deals in more detail with the problems of signal coding and transformations, and the co-ordination of transducer circuits and microprocessor programs for accepting data.

2

Signal Representation and Coding

2.1 Analogue and digital signalling

In the context of on-line computer systems it is convenient to classify electrical signals into two categories – analogue signals and digital signals.

Analogue signals are those whose amplitude conveys information about the physical quantity they represent, usually on a linear scale. In order to simplify the design of the circuits that handle these signals, some agreed full scale value is specified which the signals will never exceed. The designer must then ensure that all of the circuits will limit or overload only when the signal exceeds the agreed full scale value.

For signals that are always positive, a widely used full scale value is +10 V, giving a signal range of 0 to +10 V. This matches well with many high-gain integrated circuit amplifiers which require power supplies of +15 V and –15 V, and which limit at about ±12 V.

Where the variable can have either polarity a typical range is ±10 V. Some use is made of lower full scale voltages; for example, when connected to transistor-transistor logic (TTL) packages which operate from a +5 V supply, full scale voltages of +5 V or ±5 V may be more convenient. In rare cases when wide band signals are involved, it may be necessary to use 75 Ω or 100 Ω co-axial cable for distribution, with the cable correctly terminated at each end to avoid reflections. A full scale value of +1 V may then be used to avoid drawing excessive current from the driving circuits.

Digital signals, on the other hand, are constrained to lie only within certain voltage levels, each signal unit having only two possible values, and representing one binary digit. For example, in certain forms of logic a 0 is represented by a voltage between 0 and +0.8 V, and a 1 by voltage between +1.8 V and +3.5 V. Owing to the tolerances between different logic packages, and the effect of temperature variation, supply voltage fluctuations and circuit loading, it is not certain what signal output will be produced if the input is in the forbidden zone of +0.8 V to +1.8 V.

In order to represent an analogue signal with any precision it is necessary to specify a number of digital signals, since each digital signal denotes only two possible amplitudes. Thus using 8 binary digits we can define 256 different numerical values from 0 to 255 in decimal, or 00 to FF in hexadecimal notation. When converting a signal from analogue to digital form then, we shall constrain the analogue signal so that it is recognised as having one of only 256 different signal levels. These will be in the sequence

$$0, \frac{10}{256}\text{ V}, 2 \times \frac{10}{256}\text{ V}, 3 \times \frac{10}{256}\text{ V}, 4 \times \frac{10}{256}\text{ V, etc.}$$

or

$$0, \ 39.1\text{ mV}, \quad 78.2\text{ mV}, \quad 117.3\text{ mV}, \quad 156.4\text{ mV, etc.}$$

The most significant bit alone, as in the binary value 1000 0000 or 80 in hexadecimal, will thus correspond to an amplitude of 5 V or half full scale.

In the same way, when converting from an 8-bit digital value to an analogue voltage, the converter can generate only 256 different values corresponding to the range of hexadecimal numbers from 00 to FF.

2.2 Signal coding

This type of conversion is usually called 'unipolar binary', since the analogue signal has only one polarity, in this case positive, and the digital signal is in pure binary form.

Some analogue signals, however, are liable to have either polarity, and we may then stipulate a signal range of ±10 V. The easiest way to decode such a signal is to use a standard DAC with an output of 0 to +10 V, and follow it with an amplifier having a gain of 2 and a 10 volt offset. Thus a digit value of 00 produces an output of −10 V, a value of 80 produces 0 V, and a value of FF produces +10 V × (127/128) = +9.922 V.

This coding is called 'bipolar offset binary', and although easy to generate from a unipolar DAC, does not correspond to any code used in a digital computer.

These invariably use 2's complement coding, in which for an 8-bit number the most significant bit has a weighting of -128_{10}, the next bit $+64_{10}$, the next bit $+32_{10}$, etc. Thus the largest negative number is 1000 0000 or 80 hex, and the largest positive number is 0111 1111 or 7F, corresponding to $+127_{10}$. If now the most significant bit in an offset binary converter is negated before connecting to the DAC, the system will correctly handle 2's complement digital signals.

The various digital values in hexadecimal notation and the corresponding analogue voltages are shown for an 8-bit DAC in figure 2.1.

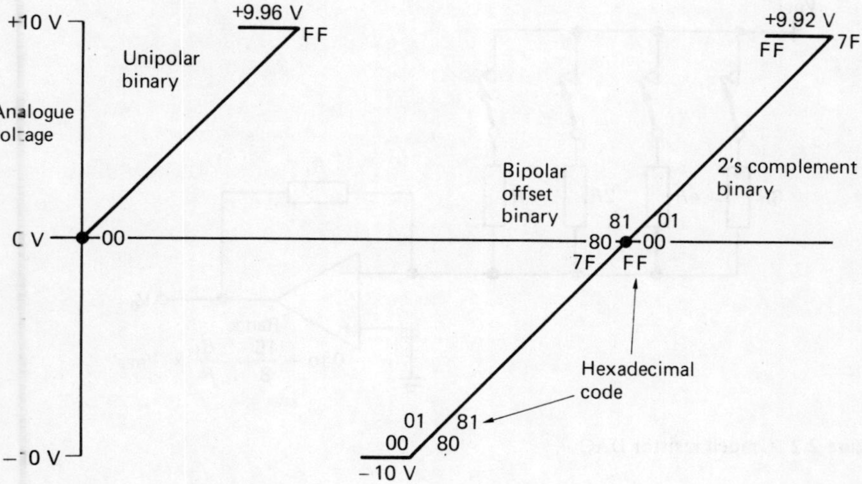

Figure 2.1 Digital coding of analogue voltage

The unipolar DAC is the simplest device to construct, and if bipolar coding is required it is followed by an offset amplifier. Similarly, when converting in the other direction, a similar amplifier with the opposite offset and a gain of $\frac{1}{2}$ is used to reduce the signal range of ±10 V, say, to the range 0 to 10 V before applying it to the ADC.

A number of packaged ADCs and DACs are fitted with amplifiers so that, by making suitable connections, either unipolar, offset binary or 2's complement coding can be adoped.

2.3 Conversion between analogue and digital signals

The simplest conversion is from digital to analogue form. This can be performed either by using a set of resistors whose values form a binary sequence, or by using a ladder network that has an attenuation of exactly 2 per stage. In either case the object is to enable each digit to switch a prescribed current into a summing junction, the currents being in a binary sequence, with the largest corresponding to the most significant bit of the number. An operational amplifier is generally used to sum the various currents and avoid interference between the digits. The switches are usually field effect transistors which are controlled by the incoming digital signals. The basic circuit is shown in figure 2.2.

Generally the ladder network is preferred since only two values of resistor are required and these can be comparatively low, of the order of 1 kΩ. Their time constant with the stray capacitance is correspondingly low and such DACs can

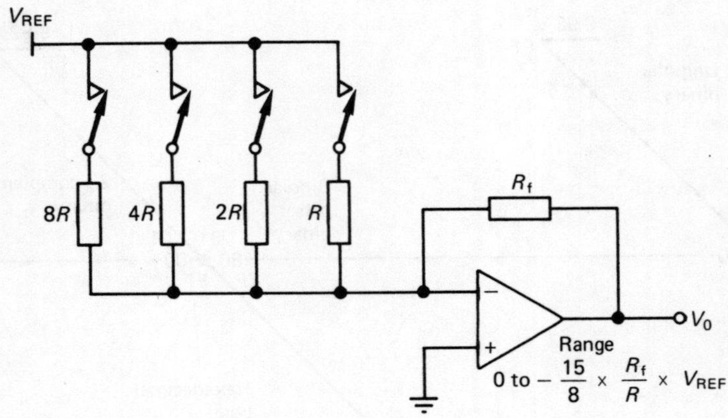

Figure 2.2 Graded resistor DAC

respond very rapidly to changing digital inputs. The general arrangement is shown in figure 2.3.

Other techniques that can be used if time allows are counter-type converters, in which the digital number is loaded into a counter which is subsequently counted down to zero. The counting time is proportional to the digital number fed in, and an integrator is allowed to generate a linear sweep during this time. At the end of the counting period the integrator output is proportional to the integration time and thus to the digital input.

This is an inexpensive technique which is adaptable for use with a microprocessor since the counting operation can be performed by the microprocessor itself, or a counter–timer package attached to it. The only external hardware needed is then an integrator, with facilities for enabling and disabling it, and also resetting it to zero.

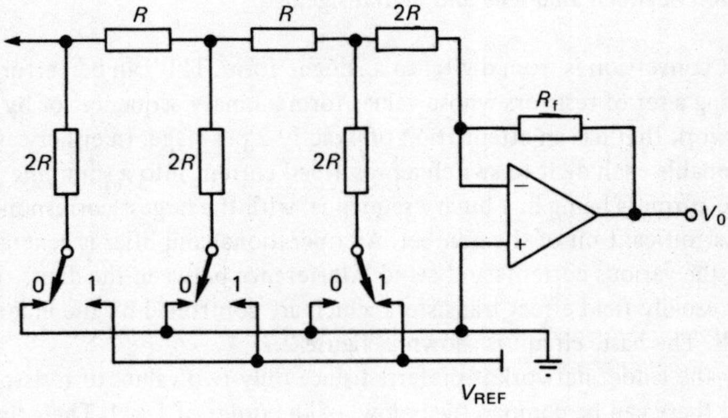

Figure 2.3 Ladder network DAC

Where only 5–6 bits provide sufficient accuracy an even simpler scheme is possible. This requires the transmission of the digital input in serial mode, with the least significant digit first. The sequence of pulses is passed to a parallel *CR* circuit, chosen so that the stored voltage decays by a factor of exactly 2 during the interval between successive digit pulses. The voltage remaining as the last pulse is applied is stored in a sample-and-hold circuit and is proportional to the digital input.

The reverse operation of analogue-to-digital conversion is a more complex operation than digital-to-analogue conversion, and generally uses a DAC together with a comparator. The two inputs to the comparator come from the DAC and the analogue input, as shown in figure 2.4.

The ADC must be designed to search for a digital input to the DAC which makes its output as nearly as possible equal to the analogue signal being converted. This input is then the digital value required and it can be fed to the microprocessor.

Several strategies are available, a quick method being that of successive approximation. In this the most significant bit is set, and all other bits cleared. The comparator output will then indicate whether this is too large or too small, when it examines the DAC output and the analogue signal. The MS bit is left as 1 if the DAC output is the smaller, otherwise it is cleared to 0. The next bit is then set to 1 and the comparison is repeated. N comparisons are required for an ADC with N-bit resolution.

The logic required for this operation, together with the register that holds the evolving output number, can be purchased as a single integrated circuit. Alternatively, the logic can be embodied in a short microprocessor program. The only hardware needed is then an I/O package or a register for holding the digital output number, a comparator, and a single digit input to the processor which carries the logical output of the comparator. This method minimises the hardware cost but is somewhat slower than using the successive approximation register package.

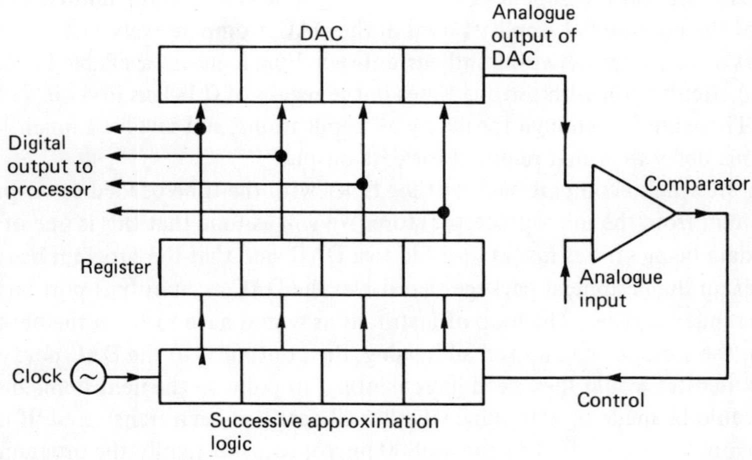

Figure 2.4 Successive approximation ADC

A simpler but slower strategy is to configure the output register as a counter and start counting up from zero. As soon as the DAC output exceeds the analogue signal the counter is stopped and its contents are sent to the microprocessor.

A number of other techniques for digital-to-analogue conversion have been used, including the so-called 'flash' converter for fast operation. This requires $2^N - 1$ comparators for an N-bit conversion, together with output logic to generate the required N-bit number. Each comparator has a reference signal in the series $\frac{1}{2}v$, $1\frac{1}{2}v$, $2\frac{1}{2}v$, $3\frac{1}{2}v$ etc. If the first three comparators indicate 'analogue signal greater', and the remainder 'DAC output greater' the required value must lie between $2\frac{1}{2}v$ and $3\frac{1}{2}v$, and will be assigned the nearest integral value of $3v$. Here v is the voltage increment corresponding to the LS bit of the digital output.

Such converters are expensive, but are necessary for converting fast signals such as those encountered in colour television or high resolution radar systems.

2.4 DAC conversion times

There is always some time interval between applying a digital input to a DAC and the appearance of a corresponding analogue output voltage. Since the output change, at least in its final stage, is generally exponential, it is usual to regard the change as finished when the output is within $\pm\frac{1}{2}$ bit of the steady-state value. The conversion time interval so defined is called the 'settling time' and is an important performance factor of DACs.

Generally the major component in the settling time is due to the limited slewing rate of the operational amplifier used in the DAC. Comparatively fast converters using ladder networks and current-switching logic are available in integrated circuit form with settling times in the region of 0.1–1 μs in 8-bit versions. These are fast enough for nearly all applications, and involve a much shorter time delay than that required for data output.

It is interesting to compare such settling times with the time needed to output one data item from the microprocessor store. We will assume that this is one of a block of data being sent as fast as possible to a DAC, and that the program has already set up the peripheral package that drives the DAC as an output port and loaded the index register. The loop of instructions would have to fetch the next item from the store, using indexed addressing, then output it to the DAC port. The index register would then need incrementing, to point to the next item, and a check would be made to determine whether all data had been transferred. If not, the loop would be repeated. For the M6800 microprocessor family the program would be

```
          Instruction              Comment
LOOP: LDA A    X,O       Load next item into ACC A
      STA A    PIO       Output to PIO port
      INX                Increment index register
      CPX END            Compare with last address in block
      BNE LOOP           Repeat if not finished
```

Here END is the address of the last data item in the store, and the address of the first item must previously have been loaded into the index register.

The program loop takes 21 clock cycles to execute, corresponding to a time of 21 μs using a 1 MHz clock. The Z-80 can perform the same task in about 15 μs using a 2.5 MHz clock. In this case the HL register can be used as a data pointer, and register B as a counter. The program would be as follows

```
          Instruction              Comment
LOOP  LD   A,    (HL)     Fetch next data item
      OUT  N              Output to port N
      INC  HL             Increment address
      DJNZ LOOP           Repeat if not finished
```

Here the B register must previously have been loaded with the number of items held in the store. It is decremented and tested in each loop and, when zero, the program leaves the loop.

The two program segments given above use continuously running program loops to output data and occupy the processor fully. If the processor has other tasks to deal with, the data rate will be reduced. Clearly with this method of data output a DAC having a settling time of around 1 μs is more than adequate, but counter-type DACs which use a ramp waveform generator while a counter decrements to zero from the input value will usually be too slow as their conversion times are measured in milliseconds.

The fastest data output rate is obtained using direct memory access (DMA). This involves disabling the processor and allowing the data store to communicate directly with the peripheral device, generally via a special DMA package.

Using the system clock to time data output will then allow an output rate of 1-2 million bytes/s, assuming a data store with a read cycle time of 250-300 ns, and operation in continuous or burst output mode. A fast DAC having a settling time of around 300 ns will be required for this arrangement, but there are comparatively few applications that need such a high data rate.

Counter DACs can be assembled using either the microprocessor itself or a counter-timer package to generate a time interval during which an external ramp generator produces the required analogue voltage. These require the minimum of additional equipment, but as the conversion time may be $\frac{1}{2}$ to 1 ms the computer program will have to restrict the output data rate to ensure that each conversion has finished before fresh data is written to the converter.

2.5 ADC interfacing

Most ADCs involve some algorithm to evaluate a digital quantity such that, when fed to a DAC, it generates an analogue voltage as near as possible to the input signal. Inevitably, therefore, they require considerably longer conversion times than DACs – 10–30 μs are typical of inexpensive 8-bit ADCs. These are now available with tri-state output buffers which can be connected directly to a microprocessor data bus. This avoids the need for a parallel interface package, and requires only small scale logic for address decoding and status signals. Their interaction with the microprocessor is, however, somewhat more complex than that of a DAC since they require a command signal to start the conversion sequence. This can be provided by one bit of an output port, or by a control line from a peripheral interface package. The program must then make provision for a delay, while the conversion occurs, before reading the contents of the ADC output buffer.

This delay can be provided in several ways, for example by using a fixed timing loop which is designed to hold up the program for a time somewhat longer than the ADC conversion time. Where time is critical, this may not be the most suitable method since the conversion time may depend to some extent upon the size of the signal to be converted. The delay must be a little longer than the greatest conversion time.

The waiting time can be minimised if the delay loop can be ended as soon as conversion has finished. In order to permit this, ADCs usually have a 'busy' signal which is at logic 1 level during conversion, and at 0 level at other times. Instead of decrementing a register, and stopping when the contents have fallen to zero, as we do in a delay loop, we can repeatedly test the busy line, and end the loop when it falls from logic 1 to 0 level. Following this, the data is read from the output of the ADC into the microprocessor. We also need to generate a short pulse to start the conversion process.

For simplicity, we assume that the digital output contains no more than 8 bits, so that it can be read in and stored as one byte in a typical 8-bit microprocessor.

The sequence of operations is as follows

Output logic 1 to 'start conversion' pin on ADC
Output logic 0 to 'start conversion' pin on ADC
Test status line
Repeat if status line is 1
otherwise, conversion is finished, read data into microprocessor accumulator.

Here the start conversion input to the ADC is assumed to require a short positive-going pulse, and the status line is assumed to have logic 1 level during conversion, and logic 0 level at other times. The timing of the various signals is shown in figure 2.5. The above program segment, of course, only commands the data conversion and reads the result into the microprocessor accumulator. Should the program require a number of readings of data, it will also need to deposit the data into the store, increment the address pointer, and test whether the number

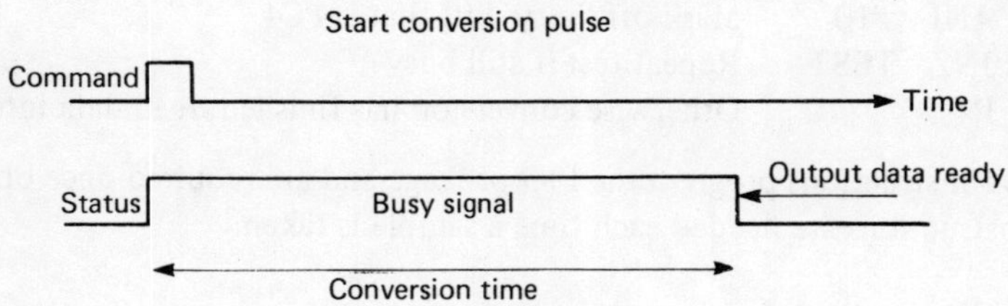

Figure 2.5 Typical timing sequence for ADC

of items required has yet been read. For some applications, readings may be required at fixed time intervals, in which case after each item has been read and put into the store the program will need to enter a waiting loop until the next timing pulse arrives.

To illustrate the program activity required, the following segment deals with an ADC interfaced to a 8080 or 8085 microprocessor via an 8255 PIO package. This has two 8-bit data ports and a split Port C which we use for command and status signals. Port A is used for data input, the lower half of Port C is used as output, bit PC0 being connected to the 'Start Conversion' line of the ADC. The upper half of PC is used as input, PC4 being connected to the ADC busy line.

The ports are programmed as follows

PA = input, for data
PB = input
PC lower half = output
PC higher half = input
Mode of PA, PB, PC = 0 (basic input/output).

This requires a control byte of 10011000 or 98 (hex).

The port addresses are as follows

Port A = 10
Port B = 11
Port C = 12
Control Register = 13.

The program required is

	Instruction		*Comment*
	MVI A,	#98	Load A with control byte
	OUT	13	Output to control register
	MVIA,	#01	Load A to set 'start convert' line
	OUT	12	Output to Port C to start conversion
	MVIA,	#00	Clear A
	OUT	12	Output to Port C to end start conversion pulse
TEST:	IN	13	Read Port C

```
ANI  #10    Mask off 'busy' bit, that is PC4
JNZ  TEST   Repeat test if still busy
IN     10   Otherwise conversion has finished, read data into A
```

The first two instructions program the PIO package and are required once only; the other instructions are needed each time a sample is taken.

2.6 Use of program interrupts

The unproductive processor time spent in waiting loops can be avoided if necessary by arranging for the end of the busy signal to create an interrupt. The main program then need only generate a 'start conversion' pulse and continue with other tasks. The interrupt service routine has no need to test the busy line since it can be entered only after conversion has finished. The service routine need only read the data into the accumulator and, if necessary, transfer it to a dedicated area of the store.

If samples of data are required at fixed time intervals, no action in the main program is called for. The timing pulses are used to start conversion, and when conversion has finished the interrupt routine will read in the data. The only action needed in the main program will be to set up the PIO control register correctly, and to enable interrupts. As an example we take an 8-bit ADC interfaced to a M6800 microprocessor via a 6821 peripheral interface package. We use Port A (PA) as the input port connected to the ADC output register, and use control line CA1 to sense the negative-going edge at the end of the busy signal. The second control line CA2 is used as an output line to start conversion. If we assume that the system has just been initialised with all interface registers cleared, the data direction registers will be set up for PA and PB as inputs. Thus we need only write information to control register CRA. The B port may also need programming for some other device, but here we consider only the A port to which the ADC is connected. CRA must be programmed for the following operations

.CA1 is enabled to create interrupts.
.CA1 flag must be set on the high-to-low transition of the busy signal.
.CA2 must be programmed as an output line, used to start conversion.
.CA2 must be programmed to copy the data written to control register bit CRA3.
.CRA2 must be set to 1, so ensuring that subsequent data transfers from Port A read the ADC output.

Two bytes must be written to the control register; the first has CRA3 = 1, so setting CA2 to 1 to start conversion. The second byte will be identical except that CRA3 will be zero, so ending the start conversion pulse generated by CA2.

The two bytes required for CRA are

0011 1101 or 3D to start conversion pulse
0011 0101 or 35 to end conversion pulse.

CRA6 and CRA7 are read-only bits and are not affected by any write instruction. By convention they are written as 0.

We assume the following interface addresses

8004	PA and DDRA	Data – Port A
8005	CRA	Control and status A
8006	PB and DDRB	Data – Port B
8007	CRB	Control and status B

The main program must write 2 bytes to the control register CRA to generate the start conversion signal, and then enable interrupts. All subsequent operations are programmed in the interrupt service routine.

The essential instructions are

Instruction	*Comment*
LDA A #3D	Load bit pattern into ACC A
STA A 8005	Output to control register A to start conversion pulse
LDA A #35	Load bit pattern into ACC A
STA A 8005	Output to control register CRA to end conversion pulse
CLI	Enable interrupts

This program segment is required for each conversion, to generate the start conversion pulse.

The service routine below copies the data from PA and resets the flag for CA1 which created the interrupt. Interrupts will be enabled after the next start conversion pulse has been generated.

Instruction	*Comment*
LDA A 8004	Read output of ADC
RTI	Return from interrupt

This leaves data in Accumulator A to be stored away for subsequent processing.

The program to be executed for each data conversion is significantly smaller and takes less time than that using a waiting loop. It is thus the preferred method when processor time is critical. However, many microprocessor applications involve external actions whose timing is dictated by mechanical operations that occur at a much slower rate than the computer can execute instructions. For these, program-controlled transfers without interrupts are entirely adequate, and are generally used as they need less hardware.

2.7 Sample-and-hold circuits

Successive approximation ADCs can work correctly only if the analogue voltage does not change by more than 1 bit during the conversion process. This imposes a severe limitation on the highest frequency that can be converted accurately, particularly with high precision converters, and where this problem could introduce serious errors it is customary to precede the ADC by a sample-and-hold unit. This takes a short sample of the analogue voltage and then holds this value constant during the conversion process. Somewhat more complex control facilities are then required since the sampling pulse must precede the start conversion pulse. A single command signal from the microprocessor is usually adequate. This operates the sample-and-hold circuit and an astable multivibrator can be used to generate from it a start conversion pulse a little later. Some packaged ADCs have a sample-and-hold circuit incorporated and require only a single pulse to inititate the sequence of operations.

Some types of ADC do not require a sample-and-hold circuit since their conversion time is effectively quite small. The 'flash' or parallel encoder is one of these, since its conversion essentially involves taking a very short sample of the analogue voltage and simultaneously comparing it with analogue voltages representing all possible digital outputs. Fast logic circuits generate the required output value from the logic outputs of all the comparators. Clearly, to obtain even moderate resolution, such as 6 bits, needs 64 comparators and a comparable amount of logic. Thus this technique, although very fast, is restricted to specialised applications on account of cost.

Another type of converter which normally responds within a computer clock cycle is the tracking converter. This resembles the counter–DAC type of converter, which counts up from zero for each conversion until the DAC output exceeds the analogue input, then stops. The tracking counter runs continuously, but the counter has an up/down input which enables it to count up or down. This is connected to the output of the comparator, so that, if the DAC output is below the analogue input, counting is upwards. As soon as the DAC output is above the analogue input, the counter reverses and counts down. The counter thus tracks the analogue input and, if correctly designed, its digital output should never diverge from the correct value by more than 1 bit. For a constant input the counter normally hunts ±1 bit about the correct value.

Since the counter is continuously tracking the changing analogue input, it can be sampled at any time by the microprocessor and effectively has zero conversion time. It is convenient to relate the counter clock to the microprocessor clock so that the counter cannot be sampled just as it is changing its state.

2.8 Sensing switch action

A widely needed arrangement in microprocessor controls is to count the number of times a switch or set of contacts is closed. One way of detecting the switch state is to connect the input line via a resistor of the order of 10 kΩ to the positive supply line. The switch leads are connected to earth and to the input line, so that closing the switch changes the logic input from 1 to 0.

Initially we expect the logic value 1; we recognise a closure as soon as the state changes to 0 but, before any further closure is counted, the logic level must return to 0. Thus we count in effect the number of 1 to 0 transitions. This can be done by continually reading the logic level, usually as one bit in an 8-bit word, isolating the switch bit and repeating the reading so long as the value is 1.

If, for example, the switch is connected to bit 2, Port A of an interface package in an M6800 system, the program could be

```
          Instruction       Comment
          CLR B             Clear ACC B to act as counter
LOOP 1:   LDA A 8004        Read byte from data port
          AND A #04         Mask out all but bit 2
          BNE LOOP1         Repeat if bit 2 = 1
          INC B             Increment count
          CMPB MAX          Check if key action is finished
          BEQ END           Yes, exit program
LOOP 2:   LDA A 8004        Read byte from data port
          AND A #04         Mask out all but bit 2
          BEQ LOOP 2        Repeat if bit 2 = 0
          BRA LOOP 1        Return to loop 1
```

Here Accumulator B is used to count the number of key depressions, and the program branches to a label END when the count reaches MAX.

Alternatively, a timer could be started when the program begins, and a test instruction included in place of the CMP B MAX instruction which would check whether the allotted time had ended. This enables the program to count how many closures have occurred in a given time.

Another way to achieve the same result would be to use a control line to set one of the flags in any interface package, and continuously test the flag. The flag can be arranged to be set on the 1 to 0 transition and, as soon as a 0 is detected, the flag must be cleared, the count incremented and the flag-testing resumed. The logic in the peripheral package is designed so that the input must return to 1, that is the switch must open before the next 1 to 0 transition can again set the flag bit to denote another switch closure.

2.9 Coping with switch bounce

Either of the above arrangements is logically correct, and appears to count switch closures correctly. Unfortunately in practice both are unsatisfactory on account of the behaviour of most switches. Since these generally have some degree of toggle action to provide rapid and positive closure, the contacts meet with a considerable speed and usually bounce several times before finally settling in contact. Since the computer programs outlined can be executed in only a few microseconds, multiple closures will be recorded for each switch operation, as the bouncing phase occupies several milliseconds.

Two methods are widely used to avoid this problem, one using hardware, and the other software. The hardware solution requires a change-over contact, and a pair of gates, cross-connected to form a bistable, and is thus more expensive than a software solution which requires only a few extra instructions. The preferred solution needs only a single pole switch, and involves inserting a delay into the computer program immediately after the detection of a switch closure. The delay is made somewhat longer than the time that the switch is likely to continue bouncing, and is typically 10–20 ms. The delay is usually produced by loading some numerical value into a processor register, and decrementing it down to zero. For example, a loop comprising a decrement and a branch if not zero instruction on a M6800 microprocessor requires 8 clock cycles, and thus takes 8 μs with a 1 MHz clock. This timing is based upon using the index register which, accommodating 16 bits, allows much longer times to be generated than if an 8-bit accumulator were used. The timing means a count of 125_{10} is required for a millisecond, and a count of 1250_{10} gives a 10 ms delay. Rounding up in hexadecimal, 0500 will give a delay just above 10 ms. Thus the time delay subroutine could be

	Instruction	*Comment*
DEL :	LDX #0500	Load index register
LOOP:	DEX	Decrement count
	BNE LOOP	Repeat if not yet zero
	RTS	Return from subroutine

This delay subroutine should be called from the main program immediately a switch closure is detected, so that the testing loop cannot be entered until the contacts have stopped bouncing.

This software procedure is typically used for sensing closure on any push-button or key that is operated manually. It can be used also when sensing closures of a reed relay, but since we have introduced a 10 ms delay we cannot handle an input that generates switch closures at a greater rate than say 80 per second. These must either come from a switch having a considerably lower bouncing time, or a change-over switch followed by a bistable must be used.

2.10 Digital signal levels

Although nearly all microprocessors are fabricated in MOS logic, so much external logic and equipment uses transistor–transistor logic that most peripheral packages are specified as 'TTL compatible', meaning that they can drive and be driven by TTL packages. Since the MOS logic usually has lower current capacity than TTL, it can typically drive only one or two standard 7400 gates, or four 7400LS gates. Typical signal levels for microprocessor interface packages are 0.8 V or less for a logic 0 and 2.0 V or more for logic 1. Some microprocessor buses also adopt these signal levels, but not all. Also most clock signals require a greater amplitude, typically 4.4 V for a system having a 5 V power supply.

On occasions it is necessary to drive CMOS logic packages of the 4000 series from TTL devices, and special steps are needed on account of the different signal levels. CMOS circuits are symmetrical, using both *p*-channel and *n*-channel devices, and the input signals need to fall to 1.5 V or less for 0 and rise to 3.5 V or more for a 1, when operating from 5 V power supplies.

Normal TTL gates which have two series transistors in a 'totem-pole' connection in the output stage may produce an output of only +2.8 V when loaded by other similar gates. This level is insufficient to drive properly a CMOS gate connected to the same +5 V supply line. However, a few TTL gates are available which have only the lower of the two output transistors included, and these are able to drive CMOS properly.

They require a 'pull-up' resistor to be connected between their output and the +5 V line, so that when generating a logic 1 the one output transistor is cut off and the output is raised nearly to the +5 V level. The resistor value is generally of the order of 1 kΩ, but this can be increased to, say, 10 kΩ if maximum response speed is not required. These TTL gates are called 'open-collector' gates – examples are

7401
7405
7409
7412
7433 etc.

Since TTL inputs can be taken up to +5 V, there is no conflict of signal levels when driving TTL gates from CMOS. The only other requirement is that the CMOS gate can sink enough current to drive the TTL input to a suitable 0 level. The current involved may be as much as 1.6 mA per input for standard 7000 TTL and 0.36 mA per input for 7000LS gates.

Another incompatibility sometimes arises because CMOS logic can operate satisfactorily with any supply voltage from +3 V to +18 V. CMOS systems are often driven from 12 V storage batteries, particularly when used in vehicles or in portable equipment. The large signal swing gives good noise immunity, but it is necessary to limit all signals that drive TTL packages so that the positive excursion does not exceed 5 V. This can be done by a small series resistor and a diode connected between the TTL input and the +5 V line.

2.11 Avoiding noise and interference

A major distinction between analogue and digital signals is their resistance to extraneous signals induced from nearby power lines and apparatus.

Analogue circuits are much more susceptible, since any noise voltage added to the signal and having an amplitude greater than that corresponding to 1 bit will introduce an error. Thus the higher the resolution the lower the permissible noise level. For example, a moderate resolution 8-bit system with a full scale signal of 5 V cannot tolerate a noise voltage in excess of 20 mV.

A higher resolution 12-bit system having the same range will suffer errors unless the noise level is reduced to around 1 mV.

By comparison a typical digital system can tolerate a noise voltage of 0.6 V before a 0 is incorrectly recognised as a 1. It is usual to base design upon the 0 signal level since the noise margin in this case is less than that permitted for a 1 signal.

The great disparity between the permissible noise levels in analogue and digital circuits emphasises the need for very careful screening and isolation of all analogue circuits.

Unfortunately many transducers have of necessity to be incorporated in all types of plant, machinery and equipment and may require long cables to connect them to the microprocessor, so affording ample opportunity for picking up noise.

The noise may arise from either electrostatic or magnetic coupling. Electrostatic coupling may be reduced by screening all signal leads with a conducting earthed shield. Magnetic screening is more difficult, particularly at low frequencies, and the usual practice to reduce interference from magnetic fields is to use tightly twisted pairs of conductors operating in a balanced mode. The object is to ensure that both wires have similar induced voltages which cancel out in the circuit loop.

One difficulty that sometimes arises is that of multiple earths. When a transducer is located at a considerable distance from the microprocessor it may be desirable in the interests of safety to connect it to the local earth. In an industrial environment this may be at a somewhat different potential from that of the earth at the microprocessor, and unless care is taken this difference voltage may be injected into the signal circuit, causing gross errors. Some rejection of this type of noise is possible by using well-balanced circuits with amplifiers designed to reject it in favour of the signal, but a more satisfactory solution is to avoid any direct connection between the remote transducer and the microprocessor.

This can be effected most conveniently by using light as a medium of information, so allowing the electrical circuits on either side of the optical link to be at quite different potentials.

The devices for this are called 'optical isolators' and comprise a light emitting diode and a phototransistor mounted in transparent plastic material close together, so that they have good optical coupling but the insulation between them will withstand typically 3 kV d.c. High voltage versions are, however, available which will withstand 10 kV. Direct capacitance between input and output is of the order of 0.5 pF.

The data rate they can handle varies from 300 k bits/s to 10 M bits/s. Their current transfer ratio (the ratio of output current to input current) varies from 20 per cent to 700 per cent for isolators that have some gain in the output circuit. The input current needed for satisfactory operation varies from 0.5 mA to 16 mA, depending upon the amount of gain available. Many versions of isolator are assembled as two-channel packages in an 8-pin dual-in-line form, and in some cases four-channel packages in 16-pin format.

In a very hostile environment where large magnetic fields cause a great deal of noise pickup, it may be worth dispensing with electrical transmission of signals in favour of optical fibres. These initially cost more than conventional screened signal cable, but are completely immune from interference caused by large stray magnetic or electric fields. They are particularly useful in devices such as particle accelerators where the transducers may be at a potential of many hundreds of kV above earth. This is a potential drop that can easily be withstood by a few metres of optical fibre, but is quite beyond the capabilities of optical isolators.

Another reason for using optical isolators arises in microprocessor systems connected to the mains supply; for example, in power control applications the microprocessor could be used to generate a firing pulse for a triac or thyristor. For this the timing reference is usually obtained from the point when the mains supply waveform crosses the zero voltage axis.

Safety regulations normally require any apparatus connected to the mains to withstand a potential difference of 1.5–2 kV a.c. between the mains leads and the equipment earth. This is provided in the power supply by using a double-wound transformer for the mains input, and adequate isolation for the zero-crossing detector can be provided either by driving it from a winding on the power supply transformer or from its own double-wound transformer.

Output circuits that drive triacs or thyristors connected to the mains must be isolated either by optical isolators or by well-insulated pulse transformers.

Where operating speed is not important and closure times of some milliseconds are acceptable, relays may be used for isolation. They have the advantage for output circuits that several banks of switch contacts can be operated by one actuating coil.

In addition to the use of screening, noise pickup in transmission lines can generally be reduced by minimising the bandwidth of the line so that it can handle the range of frequencies present in the signal, but no more. Thus it is often useful to include a low pass filter in series with the signal leads; this is particularly useful to diminish impulsive noise which has a wide bandwidth.

Another scheme for reducing capacitatively coupled interference is to lower the source resistance of the circuit driving the cable, either by using emitter follower drivers, or operational amplifiers connected as voltage followers.

3

Microprocessor Interfaces

3.1 The use of programmable packages

Although transducers and other data handling devices have been attached to digital computers almost since their inception, the introduction of large scale integrated circuits caused considerable changes in the design of interface circuits.

These were mainly caused by the change in manufacturing techniques. In minicomputer systems most interface boards were arranged to plug into the computer bus and were built in moderately sized batches from small scale integrated circuits. In popular minicomputer systems, the manufacturers' catalogues list dozens of different boards designed to drive a wide range of peripheral devices. This variety is economically justified by the nature of the assembly process.

However, once storage and central processor functions have been committed to silicon, the customer expects similar low cost and integrated circuit fabrication for the interface circuits. In view of the very high cost of initial design and mask making, it would be far too expensive to produce the very large variety of interface circuits which were previously built in printed circuit form.

The manufacturer is thus forced to reduce as far as possible the number of interface packages designed to attach to each microprocessor family. The only way in which the wide variety of applications can be catered for is then to make the interface as flexible as possible, usually by requiring some control registers to be loaded before the interface package can handle data. This is in marked contrast to the specialised minicomputer interface boards, which often need some hardware connections to set up particular address decoders etc., but thereafter are instantly ready for use as soon as the power is applied.

Most microprocessor interface packages have only volatile storage, so that, each time power is applied, they must be programmed before they can handle any data. This takes only a few instructions – perhaps a dozen or two – and 10-30 μs, and is a very small price to pay for the flexibility it permits.

Early microprocessor systems, apart from program and data storage packages, and clock circuits, provided only two or three interface packages. The three usually available were for parallel data transfer, serial data transfer, and a counter-timer function. As the market expanded other packages were added, for example for controlling DMA transfers, arbitrating interrupt priorities and controlling floppy discs.

To illustrate the range of options that may be set up by the initial program, a typical parallel I/O package having two 8-bit data ports and two control lines per port requires two registers to be loaded to control each port. The contents of these decide

Whether interrupts are permitted
Whether each of the 8 port lines is an input or an output
Whether the interrupt flag is set by a positive-going or a negative-going transition
Whether the second control line is an input or an output
Access to either the data port or one control register.

For completeness we should also mention the ultimate in programmable interface packages – a device that includes an 8-bit microprocessor having small program and data stores and two external data ports. Such a device can handle data or code conversion and other simple operations, so reducing the burden on the central microprocessor.

The transfer of data between the interface package and the microprocessor is almost invariably performed a byte at a time, using an 8-bit data port connected to the external device and 8 parallel data lines connected to the microprocessor.

3.2 Input/output instructions

Early microprocessors, like early minicomputers, have special input and output instructions which transfer data between the accumulator and the peripheral device. Thus the 8080, 8085 and Z-80 have instructions of the form

OUT N

This instruction copies the data in the accumulator out to the data lines, and issues timing signals for the interface – in this case $\overline{\text{I/OW}}$. This denotes a write operation from the processor to the peripheral and it is used in the interface package as a clock signal to latch data from the data lines into an 8-bit register in the peripheral package.

The number N is an 8-bit port number which is used to select a particular register or port in the interface package. For example, in the 8255 parallel I/O package the two select lines A0 and A1 are normally connected to the two lowest address lines, so giving four adjacent port numbers to which the package responds. Unless the system is a very small one, some decoding of higher address lines is needed to distinguish between the various I/O packages, using the chip select $\overline{\text{cs}}$

pin. If, for example, address line A2 were connected via an inverter to the $\overline{\text{cs}}$ pin, the four port numbers would be 4, 5, 6, and 7. Output operations to Ports 4, 5 and 6 load data to output Ports A, B and C; output to Port 7 loads the control register.

The port number N in the instruction is transmitted on the lower 8 of the 16 address lines, numbered A0–A7. In the same way, the instruction

IN N

copies data from the data lines into the microprocessor accumulator and transmits the port number N on the lower 8 address lines. The timing signal $\overline{\text{I/OR}}$ is issued to inform the peripheral package when to transmit the data to the data lines, and may, for example, be used to turn on tri-state drivers which transfer data from the peripheral to the data lines.

This arrangement provides sufficient port numbers for any likely system (a total of 256) but is somewhat inflexible in that the only operations permitted are transfers to or from the accumulator. Thus although the instruction set may permit a wide range of operations to be performed upon data read from the store, none of these can be performed upon data read from a peripheral device until it has first been transferred to the accumulator.

An alternative mode of operation was developed for minicomputers, and adopted for some microprocessor families such as the 6800 series. In these, all peripheral registers are allocated addresses in the total storage space available for program and data storage. Most 8-bit microprocessors are now designed to address directly 64k bytes of storage, so that the dedication of a few of these to interface packages does not materially reduce the space left for other storage.

The advantage of this scheme is that the control, status and data registers in the peripheral packages can have the full set of operations permitted by the instruction set performed upon them. These for example may include

- add to accumulator
- and with accumulator
- clear
- compare with accumulator
- complement
- negate
- decrement
- increment
- rotate left
- rotate right
- subtract from accumulator
- test (compare with zero).

Such a scheme gives the programmer more scope in handling data input and output, but at a small cost in hardware. Since there are typically 16 address lines shared by all devices and storage packages attached to the microprocessor, it is necessary to decode more of the address lines than in the alternative arrangement where the interface packages had in effect their own much smaller address space.

In view of the advantages, some of the earlier processors such as the 8080 may be used in this way, which is called 'memory mapped input/output'.

3.3 Parallel interface packages

These generally provide two 8-bit data ports for external devices, together with some additional control lines, which may be used in a 'handshaking' mode. In this the processor waits for notification from the device that it can accept data and then signals to it when a byte has been sent. It is customary for all output to be stored in the package; it remains available for use until the next data byte overwrites it.

A fairly simple example of a parallel interface package is the 6821 'Peripheral Interface Adapter' (PIA) for the 6800 family of microprocessors. This has two 8-bit data ports PA and PB, each line of which can be programmed as an input or an output by writing a 0 or a 1 in the corresponding bit of the Data Direction Register (DDR). Thus if the MS bit and the three LS bits were to be outputs and the rest inputs, the byte to be loaded into the DDR would be

1000 0111 or 87 hex

One of the control lines for each port is always an input, and sets a flag in the control register when the logic level is altered. Another bit in the control register determines whether the operative change is 1 to 0 or 0 to 1, and a further bit determines whether setting the flag also creates an interrupt. The flag is cleared by writing data to the port or reading data from it.

The second control line can be programmed as an input, acting in the same way, or as an output. In this case it can be pulsed either when data is transferred, or it can copy the data sent to one bit of the control register, so providing an extra programmable output line.

The instruction set of the computer allows for data transfers of one byte, so that if the state of one particular input line is needed, the contents of the complete port must be read, and the bit corresponding to the line of interest must be isolated by the program.

Two 'register select' pins on the package are usually connected to the two lowest order address lines, so giving four adjacent addresses for the package registers. As there are six internal registers accessible to the programmer, two of these on each port must share one address; these are the data register or port and the data direction register. The selection is made by setting bit 2 of the control register to 0 (DDR) or 1 (Port).

Initially, when power is applied the reset pulse that starts the microprocessor executing its built-in program also clears all registers in the peripheral packages. Thus the first byte transferred to the package at the shared address will be loaded into the Data Direction Register. Bit 2 of the control register must then be set in order to allow access to the data port.

If, for example, the shared Port/Data Direction Register Address is 8004, and the Control Register is 8005 the following program will set up the port for the MS three bits to be inputs, and the lowest five bits to be outputs, and will then set the three least significant bits, clearing the next two. The output bit pattern is 00111.

Instruction	*Comment*
LDA A #1F	Set up bit pattern for DDR
STA A 8004	Load DDR for 3 bits in, 5 out
LDA A #04	Set up bit pattern for port access
STA A 8005	Load Control Register
LDA A #07	Set up bit pattern for data output
STA A 8004	Output to data port

The three MS bits of the data port have been programmed as inputs so that they will be unaffected by the write operation; by convention a 0 is written to them.

Generally, a more complex bit pattern would be loaded into the Control Register, depending upon the way in which the two control lines are to be used, and whether program interrupts were allowed.

3.4 The 8255 Programmable Peripheral Interface

The 8255 PPI is designed for use with the 8080 and 8085 microprocessor families. It bears some resemblance to the 6821, but has 24 pins for connection to external devices compared with the 20 of the 6821. Both packages are encapsulated in 40 pin enclosures.

The 24 pins of the 8255 are divided into three 8-bit ports – A, B and C – Port C being divided into an upper half and a lower half. The four groups of pins comprising Port A, Port B, Port C Low, Port C High can be programmed to be all outputs or all inputs separately.

As with the 6821, there are more registers to access than the four addresses that can be selected. The Control Register and Bit Set/Reset registers share one address, the register being selected by bit 7 of the byte being written (both registers are write-only).

The data ports can be used in three modes. Mode 0 or basic I/O resembles the 6821 in that output is latched in the package, but inputs are not. No handshaking or interrupt facilities are provided.

In mode 1, the input is also latched, and input data must be presented to Port A or Port B, and a strobe pulse sent to the package to latch the data in. Interrupts can be generated, using two of the Port C lines, the remaining lines being available for strobes and control lines.

In mode 2, Port A alone is available for bidirectional data transfer, and 5 of Port C lines are used for status and control.

When some lines of Port C are not being used for control or status signals for the other ports, they can be individually set or cleared using the Bit Set/Reset register.

This package affords a somewhat different and rather wider range of actions than those provided by the 6821 package, but for this reason a more complex control program is required.

3.5 The Z8420 PIO

The Z8420 package is a Parallel Input/Output Controller designed for use with the Z-80 microprocessor. It has some resemblance to the 8255, but provides only two 8-bit ports, each of which has two handshaking signals, Ready and Strobe.

The Ready signal is output from the PIO to inform the peripheral that the port is ready for data transfer. The Strobe is generated by the peripheral to indicate that a data transfer has occurred.

The PIO can be used in four modes. Mode 0 programs all 8 pins of the port as outputs, mode 1 as inputs, and mode 2 allows Port A alone to be used; each bit can function both as input or output. This enables the port to be connected to a bidirectional data bus. In mode 3 each of the 8 lines of the port can be programmed as an input or an output. Also the input lines can be combined logically by an OR or AND function to generate a program interrupt.

An additional facility of the Z8420 PIO is the interrupt vector register designed for use with the processor's mode 2 interrupt handling. In this the 16-bit interrupt vector, the location that stores the starting address of the interrupt service routine, is assembled from 8 bits held in the processor I register, and the lower 8 bits generated by the interface, which must previously have been programmed into the PIO.

Since only one interrupt line is provided, some procedure for priority arbitration is needed. This is provided by the 'daisy-chain' connection of the Interrupt Enable In (IEI) and Interrupt Enable Out (IEO) lines. The package having the highest priority has IEI connected to +5 V, and its IEO connected to the IEI of the next package. This has its IEO in turn connected to the next IEI, and so on for all interface packages. Each package in the quiescent state transfers the logic 1 level at the IEI input to the IEO output, but when any package requests an interrupt by pulling the $\overline{\text{INT}}$ interrupt line down to 0 V, it breaks the chain and switches its IEO to logic 0 level, so preventing any lower priority packages from reacting to the interrupt. The highest priority package that is requesting an interrupt will have its IEI at logic 1 level, and will generate a timing pulse by gating together the $\overline{\text{IORQ}}$ and $\overline{\text{M1}}$ bus signals, and use the pulse to gate its half of the vector address on to the data lines.

This mode of interrupt handling involves additional logic in both the peripheral package and the processor, but it provides fast priority arbitration and there is only a small delay before the processor starts executing the required interrupt service routine.

3.6 Counter-timer packages

Many transducers generate trains of pulses which need to be counted to indicate position, speed, flow etc. While the microprocessor can be programmed to count a pulse train, it is not easy to count more than one at the same time, or to allow the processor to perform other actions while counting.

A particular problem arises with interrupts, since, without a complex arrangement for nesting interrupts, whenever the processor is diverted to an interrupt service routine it will not be able to continue counting pulses. To overcome the problems all microprocessor families include a counter-timer package. These packages embody several counters which can either count external events such as pulses, transmitted from transducers or, if fed from a fixed frequency pulse source, can generate fixed time delays.

A typical package is the 8253, designed for connection to the 8080 or 8085 series of microprocessors. It contains three 16-bit down counters which can operate in either BCD or binary coding. The counter can be loaded initially by program, and its contents can be read at any time. It can also be programmed to create an interrupt if the count falls to zero.

For dealing with transducer signals that need counting for a defined interval to deduce speed, flow rate, etc., one counter could be used to count pulses and another to generate the counting intervals. At the end of each interval a program interrupt could be generated. The service routine would read the counter, reset it and store the count reading for subsequent processing.

The 6840 counter-timer package designed for the 6800 microprocessor family is broadly similar to the 8253, containing three 16-bit counters, three control registers and a status register. The counters, however, operate only in binary mode.

The Z8430, the counter-timer package of the Z-80 family, is significantly different in architecture from the two devices described above. It has four independent counter-timer channels, but their capacity is 8 bits rather than 16. In order to generate sufficiently long time intervals when driven from the processor clock, they can be preceded by a pre-scaler which can be set for a count of either 16 or 256. When used as event counters the channels can be cascaded to provide 16-bit capacity.

Interrupt handling is similar to that provided for the parallel I/O package Z8420, but internal logic automatically selects the channel with the highest priority if more than one channel requests an interrupt. For this function channel 0 has highest priority and channel 3 the lowest. Only one interrupt vector need be programmed into the package, corresponding to channel 0. The remaining three vectors are assigned to the next three even numbered addresses. As with the PIO, the Interrupt Enable In and Interrupt Enable Out pins are connected in daisy-chain mode to similar pins on other interface packages.

4

Transducers for Linear and Rotary Movement

4.1 Coded plate transducers

Position measuring transducers can be classified in several ways, one of which concerns the need for initialising the system. This scheme divides transducers into those that give an absolute indication of position directly, and relative or incremental transducers which give position relative to some starting or datum position. These latter types need to be returned to the datum position to establish the zero of measurement each time the system is energised. With full computer control this initialisation may be performed automatically and may be part of an initial self-calibration process before any readings are taken.

An example of an absolute transducer is the coded plate device which can be used for both linear and rotary motion. The plate comprises a number of tracks each of which is scanned by a brush which forms an electrical contact, or by a light beam in the case of a photographic code plate.

Unfortunately the binary code universally used in microprocessors is not suitable for code plates, as it introduces ambiguities. The problem arises because the plate may stop in any position relative to the sensors, and when two or more bits change value between adjacent codes it is not possible to ensure that they will change state simultaneously. Thus if we consider a 4-bit encoder stepping up from the state 1011, the three bits may change in any order before reaching the next state 1100. The output may thus follow a sequence such as

$$\begin{array}{l} 1011 \\ \left.\begin{array}{l} 1111 \\ 1110 \end{array}\right\} \text{false indication of position} \\ 1100 \end{array}$$

Both the 1111 and the 1110 outputs give an incorrect indication of the position.

The simplest way to avoid this type of ambiguity is to use a code in which only one bit changes state between successive positions. The most widely used code of this type is the Gray code, which in 4-bit form is

0000
0001
0011
0010
0110
0111
0101
0100
1100
1101
1111
1110
1010
1011
1001
1000

Many other sequences can be constructed that have the same property.

In addition to its use as a code for linear motion, this code is classed as 'cyclic'. This means that it can also be used for determining angular position. An important feature of this application is that the last code 1000 is adjacent to the first code 0000 on a circular code plate. This causes no problems since there is only a single bit change between the first and last codes 0000 and 1000. The same feature applies to all complete Gray codes, that is with the number of different codes equal to a power of 2. For some applications this is not acceptable – for example, if we need to measure angular position to 1°, where the required position is given in degrees, the circle must be divided into 360 increments. Here, in order to avoid ambiguity a mask output is usually produced which just covers the transition from 359° to 0°. This is used to over-ride the encoder output and reduce it to an all-zero output. In this case all the zeros are generated from the same signal so that all 1 → 0 transitions must occur simultaneously.

A 3-bit shaft encoder disc using a complete Gray code is shown in figure 4.1. Usually these plates are produced photographically, and sensed by projecting a thin radial slit of light on to the plate. Behind each track a photodiode or phototransistor is mounted to provide an electrical output dependent upon the intensity of light passing through the plate.

Some shaft encoders with limited resolution, typically 8 bits, are made from metal plates with alternate conducting and insulating sectors on each track. The plate is connected to a voltage source, and a brush held in contact with each track picks up the output signal. In order to ensure a long life the current through each brush is limited to a milliampère or two.

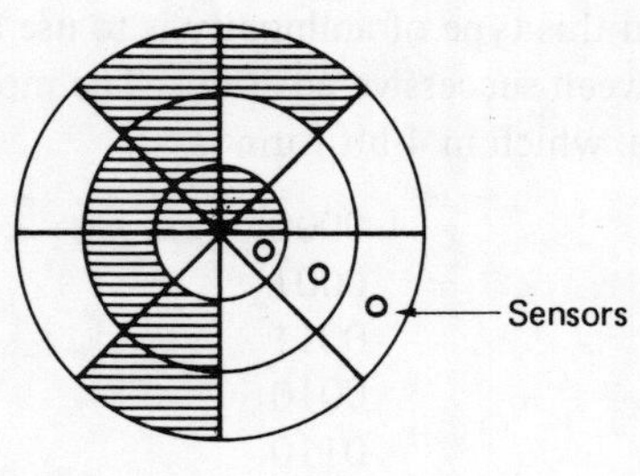

Figure 4.1 3-bit Gray-coded disc

As the Gray code is not a 'constant weight' code, that is, a 1 in any particular column does not have a fixed value, the code cannot easily be used for arithmetic. Consequently the first task after reading a Gray-coded position is to convert the Gray code into binary code. This can be done by hardware or by program.

The conversion can be performed using the exclusive-OR function which for two inputs A and B produces an output X, where

$$X = A.\overline{B} + \overline{A}.B = A \oplus B$$

If the four Gray-coded digits are G_0–G_3, and the corresponding binary digits are B_0–B_3, the logic relation between the digits is

$$B_3 = G_3 \quad \text{(most significant bit)}$$
$$B_2 = B_3 \oplus G_2$$
$$B_1 = B_2 \oplus G_1$$
$$B_0 = B_1 \oplus G_0$$

If it is important to convert the code with the minimum hardware, a set of exclusive-OR gates can be used as shown in figure 4.2. A single package contains four XOR gates, so that an 8-bit conversion requires only two DIL packages.

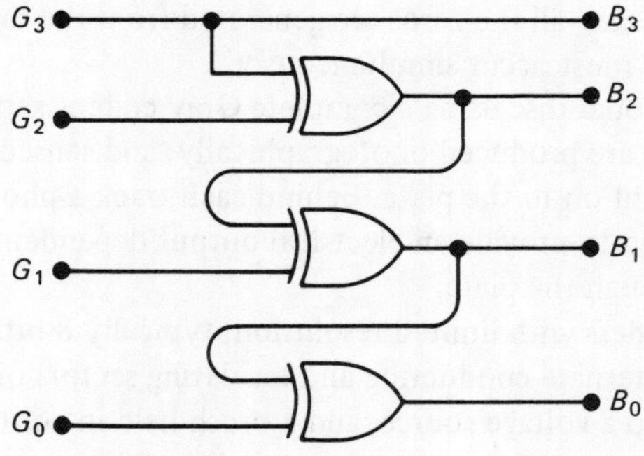

Figure 4.2 Gray to binary code conversion

Where 8 or more bits are involved an alternative conversion can be performed by a read-only store, in which the Gray code input provides the address and the binary value the stored data. Thus for a 4-bit conversion the ROM would be specified as follows

Address	*Contents*
0000	0000
0001	0001
0011	0010
0010	0011
0110	0100
0111	0101
0101	0110
0100	0111
1100	1000
1101	1001
1111	1010
1110	1011
1010	1100
1011	1101
1001	1110
1000	1111

4.2 Code conversion by look-up table

Where the conversion time is less critical the data table needed for conversion can be stored in the microprocessor read-only store and accessed by program. If we assume that the code contains no more than 8 bits, each binary value can be stored in one byte of the ROM.

As an example, using an 8080, 8085 or Z-80 microprocessor we will assume that the data table starts in location 1400 (hex). The following procedure is used

Load 14 into the H register
Read the Gray code value into the accumulator
Move accumulator contents to the L register

The HL register then contains the address of the binary code corresponding to the Gray code value held in the accumulator. This value can be read from the store by using the HL register as a data pointer, and transferred to, say, the B register for further processing by means of a move memory to register instruction. The program using 8080/8085 mnemonics is as follows

Instruction	*Comment*
MVI H, 14	Load high order byte of table address into H register
INPUT 05	Read Gray code value into accumulator
MOV L, A	Move value into L register
MOV B, M	Read data at address HL into register B

This program segment involves 28 clock cycles with an 8085 processor and will thus take 14 μs with a 2 MHz clock supply. A few microseconds could be saved if the H register is not used for any other purposes while reading Gray-coded data. The contents of the H register can then be loaded once only, and the MVI H, 14 instruction is not needed each time a reading is taken.

The program assumes that the Gray code is read from Port 5. The read-only store must be loaded as follows, for a 4-bit conversion

Address	*Contents*
1400	00
1401	01
1402	03
1403	02
1404	07
1405	06
1406	04
1407	05
1408	0F
1409	0E
140A	0C
140B	0D
140C	08
140D	09
140E	0B
140F	0A

All values are hexadecimal.

The data table is constructed in exactly the same way for all sizes of the Gray-coded input, and will contain up to 256 entries for an 8-bit code. If the Gray code value comprises more than 8 bits, two bytes will be needed to store the binary value and each entry in the read-only store will start on an even numbered address. The address will have to be loaded a little differently so that, after combining the two bytes of Gray code with a constant corresponding to the starting address of the data table, the HL register contents can be shifted left to give the address of the first byte of the binary value. The remaining part of the required binary value will be stored in the adjacent address.

Encoders that generate more than 8 bits need double-length storage – for example, a 10-bit system will need 2k bytes of storage, whereas an 8-bit system needs only 256 bytes.

Since encoders are usually available for 8, 10, 12 etc. digits, it is worth comparing costs for converting 8-bit and 10-bit codes. Using typical component costs (1 off) of £2.50 for a 2k × 8 bit EPROM and 60p for a quad exclusive-OR LSTTL package, the costs are as follows

	EPROM	*Logic*
8-bit code	256 bytes 32p	2 packages £1.20
10-bit code	2k bytes £2.50	3 packages £1.80

Thus for economy the data table method would be chosen for 8-bit converters, and the hardware conversion for higher resolution systems.

Another method could be used since most microprocessors have exclusive-OR instructions. This involves implementing the logic equations for the hardware conversion by program. It is a somewhat lengthy process since each equation must be executed separately and the bits involved must be selected by masking from the input Gray code data. This method uses much less storage space than needed for the look-up data table but will take considerably longer to execute.

4.3 Incremental shaft encoders

An alternative type of shaft encoder which gives relative angular position is the incremental shaft encoder. This requires only one set of marks on the disc in the form of equally spaced radial lines. An extra track contains a single mark which is used as a zero reference. The system is initialised by turning the shaft until the reference output occurs. This is used to clear the contents of a counter to zero. The counter subsequently increments as each bar passes the optical sensor, so keeping track of the position. Usually there are two sensors which are located so that their outputs are 90° out of phase. This enables fairly simple logic to determine in which direction the disc is moving, and so whether to count up or down. This feature is needed only if the direction of motion of the shaft can reverse – for example, if the encoder forms part of a position control system.

The signals produced by the encoders may be either rectangular waves, as shown in figure 4.3, or of approximately sinusoidal waveform.

Typical encoders may have up to 1200 lines per revolution, so giving an angular resolution of 0.3° or 18 minutes of arc. This will require an 11-bit input to the microprocessor, if a hardware counter is used to keep track of the shaft position. If the microprocessor is not subject to program interrupts, it could be used to determine position by using a software counter. Only two lines into the computer would then be needed, one counter line which produces a pulse each time a line passes the light sensor, and a direction line which indicates the direction of motion of the shaft. The latter determines whether the counting should be up or down.

The basis for a simple direction recognition circuit is shown in figure 4.3. A short pulse is generated from each positive-going edge of the $\emptyset_1$ signal, and used as

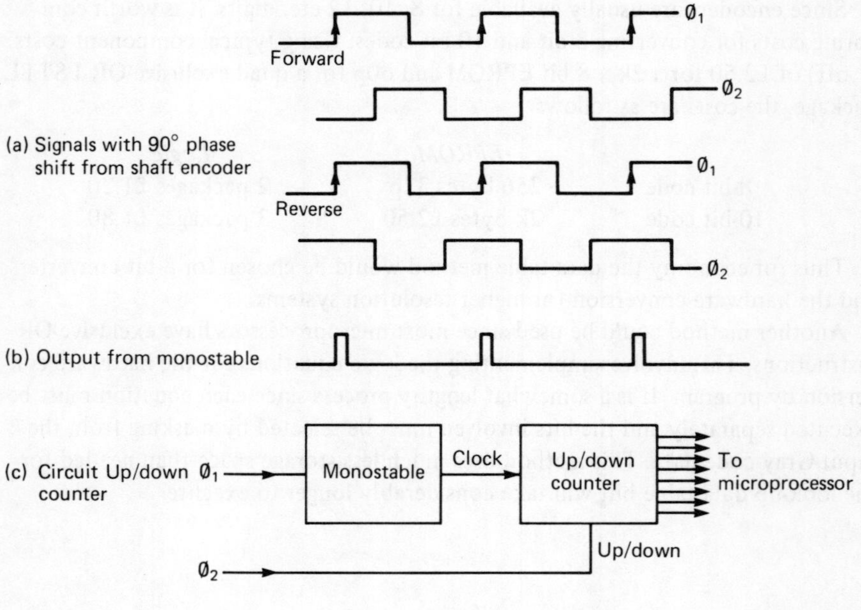

Figure 4.3 Connections from shaft encoder to up/down counter

the counter clock signal. As there is no minimum speed of rotation for a position control system, an edge-triggered monostable would be unreliable in such a scheme and it would be necessary to use either a level-triggered circuit, or to precede the monostable by a Schmitt trigger circuit which always generates a fast output transition however slow the input transition is.

The $\emptyset_2$ signal could be used as the up/down input as its value in the waveforms of figure 4.3 will be 0 for forward rotation and 1 for reverse rotation when the clock signal occurs. A typical up/down counter for this application would be the 74LS190 (decade) or 74LS191 (binary) in TTL, the 4510B (BCD) or 4516B (binary) in CMOS logic. All of these counters are 4-bit packages which can be cascaded to construct 8-bit or 12-bit systems, and can be cleared for inititalisation.

Most microprocessor families now include a counter–timer package which can count an external pulse train without the use of the central processor. These packages are normally used when program interrupts are permitted, otherwise if the processor is counting the pulses some are liable to be missed should they occur during the execution of an interrupt service routine.

The major difficulty in using counter–timer packages is that the counters normally count down only. They are thus usable only if the shaft moves in one direction without reversing. Applications would thus include synchronising the rotation of a shaft with an external timing signal, but would exclude its use in servo mechanisms that are expected to rotate in both directions.

4.4 Moiré fringe measurements

Moiré fringe measurements rely on counting methods and so determine position with respect to a fixed datum or reference position. Two transparent plates are used, each having a series of parallel dark lines marked upon it. One plate is attached to the moving object and the other is fixed, and they are arranged with their lines at a small angle. When relative motion occurs, alternate light and dark fringes move across the plates at right angles to the direction of motion.

The important feature is that the spacing of the fringes is many times greater than the spacing between the lines marked on the plates. The magnification factor is cot α, where α is the angle between the two sets of lines. When α is small and expressed in radians, $\cot \alpha \simeq 1/\alpha$. Thus if α is 0.01 radians, a little less than $\frac{1}{2}^\circ$, the magnification factor is 100. If the lines are spaced 50 per mm, so giving a resolution of 0.02 mm, the fringes will be 2 mm apart. The plates are usually made by photographing a larger original drawing. The fringes are detected by means of a narrow light beam and phototransistor. At least two phototransistors are normally used in order to detect which way the object is moving. The system is initially moved to the datum position, the counter is cleared, and the system is then ready for use. The pulses from one detector are used to derive clock pulses for an up/down counter and the pulses from the other detector can be used to control the direction of counting. The circuit is identical to that given in section 4.3 for use with incremental shaft encoders.

4.5 Inductive counting systems

Where the position of some device need not be measured in such small increments as are normally provided in moiré fringe systems, an inductively coupled arrangement can be used. The scheme is shown in figure 4.4. The upper rack is stationary, and the E-core which supports the coils is attached to the moving member. The centre coil is energised by an oscillator and the two coils on the outer limbs are connected in series opposition. The pitch or distance between the centres of the adjacent limbs of the E-core is about three-quarters of the pitch of the fixed rack. Thus when the left-hand limb A of the E-core lies opposite a tooth in the rack, as shown, the flux linking coils L_1 and L_2 will be much greater than that linking L_2 and L_3 since the reluctance of the flux path is much lower. When the E-core has moved about one-quarter of the rack pitch to the right, the centre limb B of the E-core will lie underneath a tooth of the rack, and the other two limbs A and C will both lie about half-way under a tooth.

In the first case the dominant voltage will be that induced in L_1, and the voltage induced in L_3 will be much smaller. In the second case the voltages induced in L_1 and L_3 will be smaller, but equal and of opposite phase. Consequently the nett output voltage will be very small. A further movement of a quarter pitch to the right will produce maximum voltage in L_3 and only a small voltage in L_1. Thus the

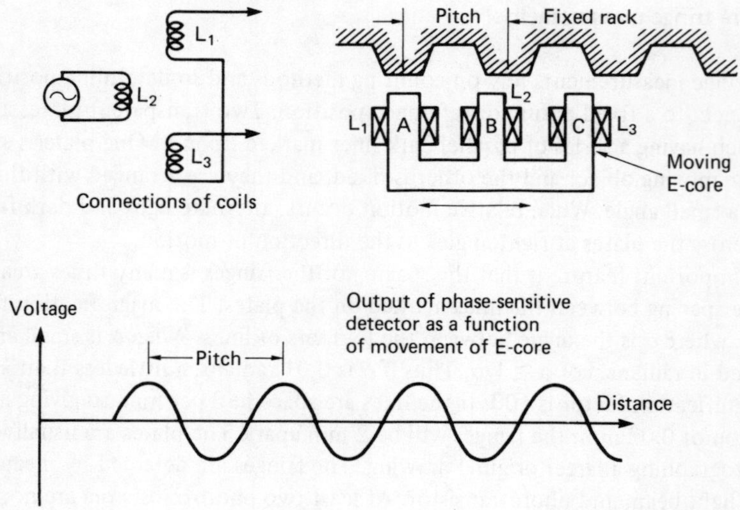

Figure 4.4 Details of inductive position sensor

phase of the output will be opposite to that generated with the E-core in its initial position.

If the output is taken to a phase-sensitive detector with the oscillator output as reference, a sinusoidal output will be produced as the E-core moves along, one complete cycle corresponding to a displacement of one pitch of the fixed rack. The most accurate positioning will be obtained by looking for zero crossings of the output voltage, so that two exact positions can be located in each pitch of the rack. A particular application of this type of transducer is to the positioning of the heads above the tracks on a magnetic disc drive as used for backing storage in computer systems. In this case, the spacing between adjacent tracks on the magnetic disc is half the spacing between adjacent teeth on the rack. This inductive positioning system is most suitable for applications where a number of equally spaced positions are required, which can be identified by the zero crossings of the output signal.

4.6 Potentiometer sensors

A simple method of measuring position is to use a linear potentiometer. This may consist of a closely wound coil of resistance wire supported on an insulating former. One end of the coil is earthed and the other end is connected to a reference potential V_{REF} as shown in figure 4.5. The slider is connected to the moving member and the coil is fixed. If the voltage on the slider is v, the displacement x from the earthed end of the coil is given by

$$x = \frac{v}{V_{REF}} \times l$$

where l is axial length of the coil.

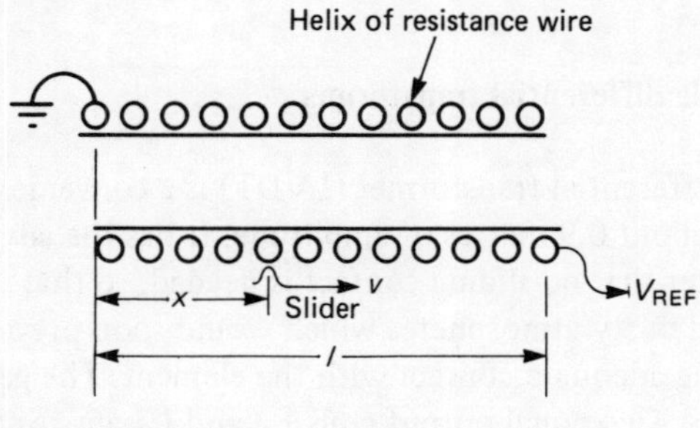

Figure 4.5 Potentiometer transducer

This technique although cheap to implement has several disadvantages. Firstly it must impose some mechanical load on the source of the movement. Unless this had sufficient power to overcome the frictional force between the slider and the resistive element the accuracy will be impaired. Secondly the output of the potentiometer is an analogue voltage which must be converted into digital form before it can be accepted by the microprocessor.

The conventional coil of resistance wire used for the potentiometer also introduces quantisation errors since the smallest step of voltage by which the output can change is that corresponding to one turn of wire. Thus if the coil contains 250 turns the resolution is 1 part in 250 or 0.4 per cent. Improved resolution can be obtained by using a continuous element such as a ring or rod of conducting plastic.

A further difficulty arises with circular motion in that some small space is needed for the end connections to the coil. Thus the output voltage is a valid representation of the slider position for perhaps 352–357° of rotation, whereas the output of a digital shaft encoder is valid for any angular position.

Typical potentiometer transducers have linearity errors in the range ±0.1 per cent to ±0.5 per cent, so that to avoid introducing further errors the ADCs used with them should have resolutions of 8–10 bits. It is also important to draw only a minimum of current from the slider, so that an isolating amplifier having a high input impedance may be required between the potentiometer and the ADC. The resolution of a rotary potentiometer may be increased and its linearity error reduced by using a multi-turn element. For example, 10-turn potentiometers are available with a quoted linearity error of ±0.03 per cent. Such a transducer will require a 12-bit converter to take full advantage of its accuracy.

In some applications involving large linear displacements, 10-turn rotary potentiometers have been used. They are fitted with a cylindrical drum and a spring which returns the shaft to the zero output position. A thin steel tape is wrapped round the drum and its free end attached to the moving object. As it moves away from the zero position the tape unwinds and turns the potentiometer shaft round. This system is convenient for the manufacturer since any required range of movement can be provided merely by changing the diameter of the drum.

4.7 The linear variable differential transformer

The linear variable differential transformer (LVDT) is a convenient transducer for linear motion up to about 0.9 metres of amplitude. It has the advantage over a resistive potentiometer that no sliding contact is needed, so that it can operate in chemically active and dusty atmospheres which would soon prevent the slider of a potentiometer making adequate contact with the element. The general arrangement is shown in figure 4.6. Two equal output coils L_1 and L_3 are connected in series opposition, and are symmetrically placed with relation to the central coil L_2. This is energised by an oscillator at a frequency of 5–10 kHz. The iron core can move freely along the axis of the coils. As it moves it varies the mutual inductance between L_2 and the output coils, and thus varies the e.m.f.s induced in L_1 and L_3.

When the core is in a central position, equal e.m.f.s are induced in the output coils L_1 and L_3, and as these are connected in series opposition, the nett output signal will be zero. As the core moves away from this central position the nett e.m.f. induced in the output coils increases in proportion to the displacement.

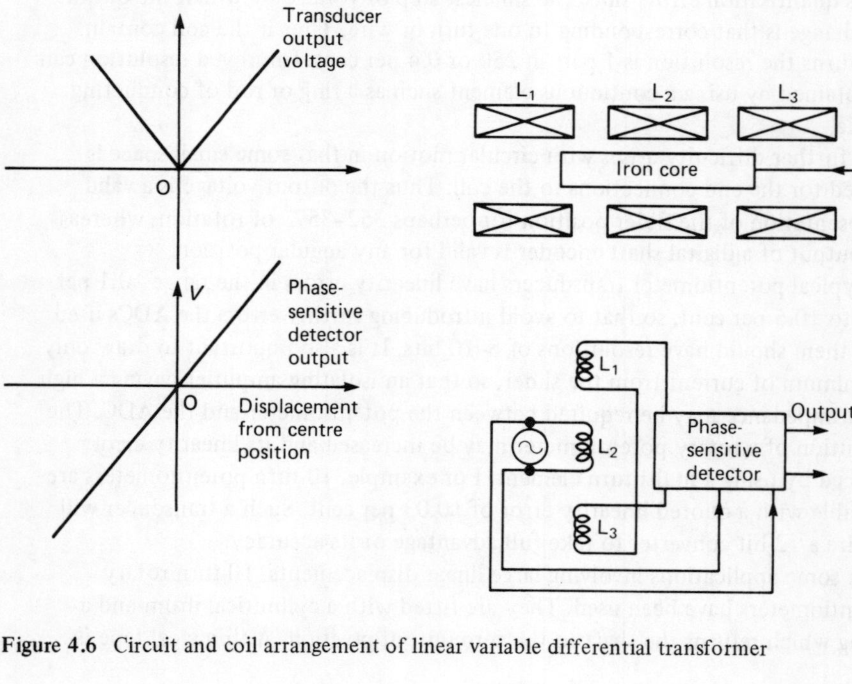

Figure 4.6 Circuit and coil arrangement of linear variable differential transformer

However, the phase of the output will depend upon the direction of movement, and the phase of the output signal will change by 180° as the core crosses the central position. Consequently, if the output is taken to a phase-sensitive detector whose reference supply is derived from the oscillator, the smoothed signal from the detector will indicate position unambiguously. It will have a magnitude proportional to the displacement from the central position, and will be negative on one side, positive on the other, as shown in figure 4.6.

The advantage of this technique is that no sliding contacts are involved, and no flexible leads are required to connect to moving coils. All three coils are fixed, and the only force involved is that needed to overcome the friction force of the bearings which constrain the moving core. This can frequently be attached to some machine member which is moving in a straight line, so that the transducer itself will not then introduce any extra frictional load.

Typical linearity errors of less than 0.5–1 per cent are quoted by manufacturers. Thus 8-bit ADCs and input ports will have sufficient resolution to capture the data provided by LVDT transducers.

For the user's convenience, many LVDT transducers are now made in what is called a 'dc–dc' configuration. This has internal electronic circuits which include an oscillator to generate the 5–10 kHz excitation signal, and a phase-sensitive detector with smoothing. The power needed is typically 6–10 V d.c. at 50 mA, and the output is a smoothed d.c. signal.

Some forms of LVDT have the coil assembly hermetically sealed so that they can be used at high temperatures and in corrosive atmospheres without damage.

For measuring small displacements a 'rocking armature' transducer which works on a similar principle can be used. In this the three coils are wound on the centre limb and the two outer limbs of an 'E'-shaped stack of laminations. The armature rocks on a central pivot, so increasing one airgap and decreasing the other. This alters the fluxes in the outer limbs, and hence the e.m.f.s induced in the two outer coils. A phase-sensitive detector is used to provide an output, as for the LVDT. The arrangement is shown in figure 4.7. It is a convenient way of measuring small displacements.

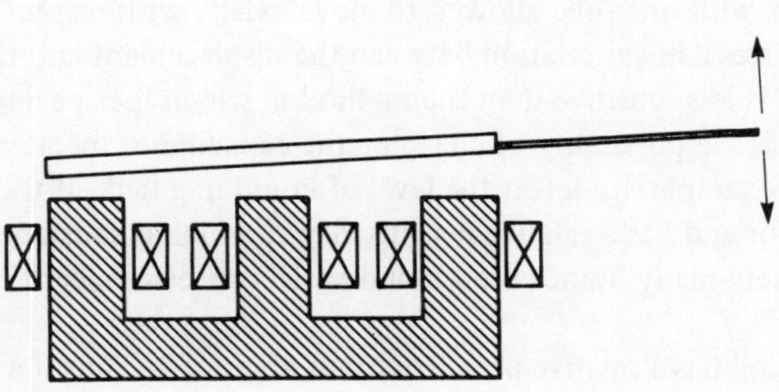

Figure 4.7 Rocking armature transducer

4.8 Capacitance sensing

The capacitance of an ideal parallel plate structure is given by the expression

$$C = \frac{\epsilon_0 \times \epsilon_r \times A}{d}$$

where ϵ_0 is the permittivity of vacuum
ϵ_r is the relative permittivity of the dielectric between the plates
A is the area of the plates
d is the spacing between the plates.

ϵ_r is almost exactly 1 for air, and if A and d are measured in metres^2 and metres, the constant $\epsilon_0 = 8.85 \times 10^{-12}$ F/metre.

The capacitance of a parallel plate system can be made to depend upon displacement in several ways, to construct a capacitance transducer. The method generally used is to vary the spacing d. The transducer is usually built into a bridge circuit, which needs only a few volts excitation. Thus the spacing between the plates can be very small, since an airgap of only 10 micrometres would sustain a potential difference of 10 V. This method of using capacitance sensing is thus most convenient for small displacements where other methods are less effective. A minor problem is that the relation between displacement and capacitance is a non-linear one. However, if d changes by no more than, say, 10 per cent, the error introduced is usually tolerable. Otherwise, if the output of the bridge feeds a microprocessor, the non-linearity may be allowed for in the computer program, either by calculation based upon the known geometry of the transducer, or by calibrating the transducer by, for example, a set of feeler gauges. A table of calibration points can then be stored in the microprocessor, and referred to after each reading.

An advantage of capacitance sensing over the LVDT is that the electrodes can be very light, consisting of thin metal foils or thin metal layers deposited on an insulator made of a ceramic material. They can thus be used on light, rapidly moving objects without imposing excessive inertial loading.

Another way of making the capacitance vary with movement is to change the area of overlap of the plates. This is often done with a transducer comprising two concentric tubes, with one tube allowed to move axially with respect to the other. This technique gives a linear relation between the displacement and the change in capacitance, but is less sensitive than the method in which the spacing is altered.

Finally there is the possibility of changing the capacitance by altering ϵ_r. This can be used, for example, to detect the level of liquid in a tank, if the liquid is an electrical insulator and has a relative permittivity ϵ_r which is substantially greater than 1. Fortunately many liquids used in industrial processes meet these requirements.

For example, oil has a relative permittivity of about 2.2. Thus if a parallel electrode system is built vertically into a storage tank, the capacitance of, say, 100 pF measured with the tank empty would increase to 220 pF when the tank was filled and the electrodes were completely immersed. This type of transducer is usually connected to a bridge circuit which is balanced with the tank empty. The out-of-balance signal is then proportional to the depth of liquid in the tank.

A similar arrangement can be used with powdered or granular solids, but the spacing between the electrodes needs to be rather larger to ensure that the powder level is properly sensed.

4.9 Ultrasonic beam sensing

Where capacitance sensing would be inconvenient – for example, if access cannot be gained to the inside of the enclosure, and it is possible only to insert some transducer through the lid of the tank – ultrasonic sensing can be used. In this scheme a short pulse of ultrasonic energy is radiated in a fairly narrow beam by a transducer placed vertically above the surface of the liquid. This is reflected from the surface and returned to the originating transducer, or a similar device placed next to it. By measuring the time interval between the transmitted pulse and the returned pulse, the distance h between the transducer and the liquid surface can be deduced. In the time interval t the sound travels a distance $2h$ at a speed of V. Thus

$$2h = V \times t \quad \text{and} \quad h = \frac{V \times t}{2}$$

For air, V is about 331 m/s, so that a time delay of 1 ms corresponds to a distance h of about 16.5 cm.

This technique can thus give more than enough resolution for most industrial and process control requirements since we can easily measure time in increments of less than 1 ms.

Depth gauges of this type are often built as self-contained systems with an analogue output to drive a meter scaled in distance. This signal can then be connected to an ADC to provide a digital signal suitable for microprocessor input.

The electronic circuits of this type of depth gauge could be made much simpler if they were designed specifically for interfacing to microprocessors. A transmit pulse could be triggered by an output line, and one channel of a counter–timer package could be used to measure the interval between the transmitted and returned pulse, by counting a pulse train of known frequency during the interval. Since the counters are 16-bit devices, the resolution of measurement is limited by the rise-time of the reflected pulses rather than the numerical capacity of the counter.

In this type of depth gauge the measured value is the clearance between the top of the tank and the surface of the liquid. This must be subtracted from the height of the tank to determine the depth of the liquid, a calculation that can easily be included in the microprocessor program.

A similar ranging technique has been used for distance measurement using very high frequency radio waves. These have the advantage that the propagation velocity is much less dependent upon air temperature, pressure and movement, but as the waves travel around a million times faster there is no possibility of using the microprocessor hardware for measuring the propagation time.

4.10 Measuring moving objects

A particular problem of length measurement occurs in many industrial processes where material is produced continuously and it is necessary to measure the length of some object as it is moving. This has been tackled in a variety of ways.

A simple procedure, if the speed of movement is known, is to make the object interrupt a beam of light or ultrasound. The time during which the beam is interrupted multiplied by the speed gives the length directly. The speed of the conveyer belt or other transport system may not be known very accurately, so that it may be necessary to measure this also if precise results are to be obtained. A conveyer belt can easily be provided with a freely moving jockey roller whose angular speed can be measured. Such a system is easy to implement if a microprocessor is used, since it can measure the time interval during which the beam is interrupted and the angular speed of the roller, and multiply them together.

For objects with a fairly smooth straight profile, a freely rotating wheel coupled to an incremental shaft encoder could be used, held in contact with the object by light spring pressure. The number of pulses emitted by the encoder during the passage of an object would then be directly proportional to its length.

Where it is not possible to attach sensors to the path of the object, image sensing has been used. This requires the object to be easily distinguished from its background, for example, a dark object against a light background. A television camera is then focused on the object, arranged so that the line scan direction corresponds to the direction in which length is to be measured. The few lines of the television raster that traverse the object are gated out from the video signal, and will in this case show a low level while scanning the object and a high level while scanning the background. The duration of the low level can be measured either by a separate counter, or by passing a train of pulses to a counter-timer package attached to a microprocessor.

A typical closed circuit television system working on the 625 line standard has a line scan time of about 60 μs. In most applications the object will move little during this time, so that the speed of the object will not introduce much error into the measurement.

5

Temperature Measurement

5.1 The platinum resistance thermometer

The most accurate method of measuring temperature in an industrial environment is based upon the relation between temperature and resistance of a coil of pure platinum wire. The wire is wound with a minimum of tension on a ceramic former and in adverse conditions may need to be protected by a sheath made of stainless steel or, for higher temperatures, ceramic.

The temperature/resistance characteristic is given in a table in *BS 1904: 1961 (1981) Industrial platinum resistance thermometer elements*. The table extends from a temperature of –220°C to +1050°C, although the upper limit is usually restricted to about 600°C. The preferred resistance element is adjusted to have a resistance of 100.00 Ω at 0°C. It will have a resistance of 10.45 Ω at –220°C, 138.50 Ω at 100°C and 446.3 Ω at +1050°C.

One of the problems in the use of the resistance thermometer is the effect of the connecting leads. These are usually of copper, and so their resistance will alter with temperature. In order to prevent this introducing errors into the measurement, compensating leads are used. These are identical to the leads used for the resistance element, but are short-circuited at the thermometer end, and connected in series with the standard comparison resistor in a bridge connection. The circuit is shown in figure 5.1. In this arrangement the cable between the bridge and the sensor has four cores, twisted together for minimum noise pickup and screened with a conducting braid. It is important to restrict the bridge supply voltage so that the current passing through the resistance coil is too low to heat it up materially and so cause an error.

The unbalance voltage is usually amplified and used to indicate temperature, or for microprocessor input connected to a DAC. Where maximum accuracy is required, the small non-linearity in the temperature/resistance relationship must

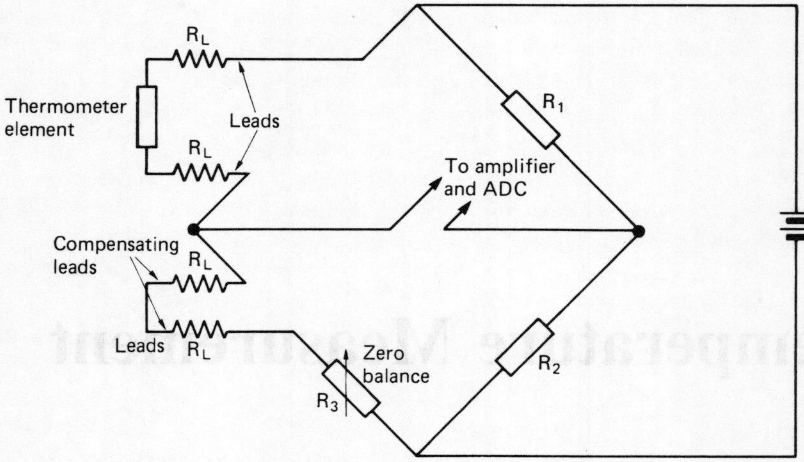

Figure 5.1 Wheatstone bridge circuit for platinum resistance thermometer

be taken into account. Expressions for this are given in BS 1904. For the range 0°C to 630°C the relationship used is

$$t = \frac{1}{\alpha}\left(\frac{R_t}{R_0} - 1\right) + \delta\left(\frac{t}{100} - 1\right)\frac{t}{100}$$

where t is the temperature in °C
$\alpha = 3.901 \times 10^{-3}$
$\delta = 1.4923$.

The correction can be made in the computer program either by using this expression or by storing a table of resistance values against temperature and interpolating between them.

In order to achieve maximum accuracy the platinum resistance element must not be subject to mechanical stress, and should be temperature cycled and aged before calibration. Under optimum conditions, the error in measurement may be as little as ±0.05°C at 500°C.

5.2 Thermocouples

The high accuracy obtainable from platinum resistance thermometers is rarely essential for industrial controls, and alternative cheaper arrangements are often used. The most important of these is the thermocouple sensor. This consists of two wires made of dissimilar metals joined together, with a break in the circuit to allow the e.m.f. to be measured. When the two junctions are held at different temperatures, an e.m.f. is generated which is almost proportional to the temperature difference. Thus if the cold junction is held at 0°C, a millivoltmeter connected across the break can be calibrated to read the temperature of the hot junction. For input to a microprocessor system the signal must be amplified by a directly coupled amplifier that has a low zero drift, before being fed to an ADC. This is necessary because the maximum output of a thermocouple is only 20–50 mV, depending upon the materials used and the temperature.

Table 5.1 gives the recommended operating temperatures, the relevant British Standard that contains tables of e.m.f. against temperature, and the maximum e.m.f. generated for various combinations of material.

Table 5.1 Thermocouple data

Standard	*Materials*	*Temp. range* (°C)	*Maximum output*
BS 1826: 1952	Pt *vs* 13 per cent Rh/Pt	0 to 1770	21.1 mV at 1770°C
BS 1827: 1952	NiAl *vs* Nichrome	−190 to 1374	55 mV at 1374°C
BS 1828: 1961	Copper *vs* Constantan	−190 to 400	20.6 mV at 400°C
BS 1829: 1962	Iron *vs* Constantan	−190 to 729	41 mV at 729°C

5.3 Thermistors and other devices

Thermocouples for use up to about 1000°C need not be made from noble metals, and are thus cheaper than platinum resistance thermometers. They do, however, require stable high gain amplifiers with well-regulated power supplies to amplify the millivolt signals to several volts before feeding to an ADC. Where a less accurate indication of temperature can be tolerated, the system cost can be reduced considerably by using a sensor that can directly generate a voltage sufficient to feed the ADC. The simplest such device is the thermistor, which has a large negative temperature coefficient of resistance, and is made from a solid semiconducting metal oxide.

The relationship between the resistances at temperatures T and T_0 is of the form

$$R_T = R_0 \exp \left\{ b \left(\frac{1}{T} - \frac{1}{T_0} \right) \right\}$$

Figure 5.2 shows this relation for a particular thermistor. The value of b is around 4000 for a reference temperature T_0 of 25°C. A typical device having a resistance of 2 kΩ at 0°C would have a resistance of only 40 Ω at 100°C. The relationship between temperature and resistance is non-linear, but it can be made roughly linear by adding shunt or series resistance.

Alternatively a table of resistance values against temperature can be stored within the computer program, and used to calculate the temperature. Where some liquid in a process has to be kept at one of several pre-determined temperatures, the program need store only data corresponding to these temperatures.

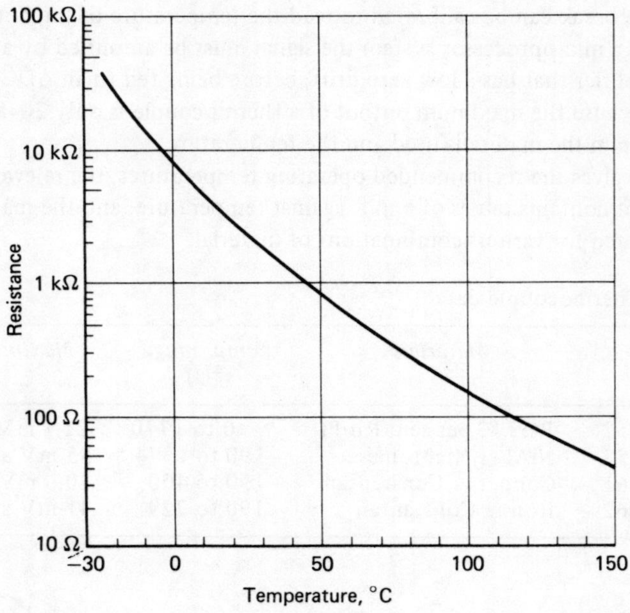

Figure 5.2 Resistance/temperature graph of thermistor

Thermistors are made in a variety of forms and with a wide range of resistance values. They are usable at temperatures up to about 300°C. Some types can be made very small, so that they can attain the temperature of their surroundings in milliseconds.

A device that offers a good sensitivity and linearity, but is more expensive than a thermistor, is the current generating temperature sensor. This device is available with an accurate calibration of 1 μA per °K, when a voltage between 4 V and 30 V is applied. It needs no cold junction compensation since the current output is zero at the fixed temperature of −273°C, but it will need a zero offset if it is to produce an output proportional to degrees Celsius.

The operating temperature range of the current generating temperature sensor is −55°C to +150°C. When used for microprocessor input it is not essential to introduce a zero shift as this can be allowed for in the computer program. However, this method sacrifices some of the dynamic range available since the temperature range −273°C to −55°C cannot be used. This means that the working range of −55°C to +150°C corresponds to about half of the total range of −273°C to +150°C. When used with a 10 kΩ load resistor, the output voltage is 10 mV per °K, or 1 volt for a 100°C temperature change. At 0°C it will produce 2.73 V, and 3.73 V at 100°C.

A convenient arrangement would be to feed this signal into an ADC having the same scaling of 10 mV per bit. This would then give an output of 1 bit per °C. This resolution is sufficient for many purposes, and could be obtained by using the ZN425 or ZN427 8-bit ADC which has a scale factor of 10 mV per bit and a maximum input of 2.55 V. For greater resolution a 10-bit or 12-bit ADC could be used. The input impedance of the ADC is too low for direct coupling, so that an operational amplifier could be used for scaling and offset, and for providing a low impedance source for the ADC. A suitable circuit is shown in figure 5.3. The zero set potentiometer should be initially adjusted to provide about 2.73 V to the 10 kΩ resistor connected to the slider. The final adjustment should give zero output at 0°C.

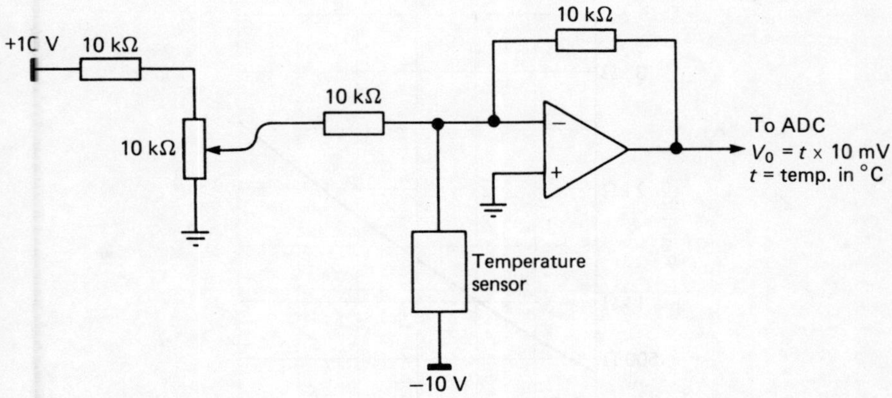

Figure 5.3 Zero offset and buffer circuit for temperature sensor

Temperature can also be measured by observing the change in characteristics of a transistor with temperature. The usual arrangement is to operate a transistor at constant collector current. It is then found that the base–emitter voltage falls by about 2 mV for every °C rise in junction temperature. This is much greater than the 50 μV or so that the most sensitive thermocouple generates for a 1°C temperature change.

The practical difficulty is that the base–emitter voltage at 0°C is around 0.7 V, so that there is a large standing voltage to be offset when using this arrangement. Consequently stable power supplies and a low-drift, high-gain amplifier are required, and the overall cost may exceed that involved in other methods of measurement, in spite of the very low cost of the transistor used as a sensor.

Thermistors and transistors have a negative temperature coefficient – that is, their resistance, or their base–emitter voltage, falls as the temperature rises. Another type of device having a positive temperature coefficient has recently been introduced, the silicon temperature sensor. This uses *n*-type silicon and can be manufactured with much smaller tolerances on its characteristics than can thermistors. The relationship between temperature and resistance is less non-linear than that of the thermistor, and it can be made nearly linear by adding a resistor in parallel. Alternatively, if a resistor is added in series, and the combination is fed from a constant voltage source, the voltage across the sensor is an almost linear function of temperature. This leads to a simple arrangement which needs only a small offset voltage to yield an output proportional to temperature in degrees Celsius.

As with other devices, the self-heating of the sensor due to the measurement current can introduce errors. These are negligible if the current does not exceed 1 mA for the sensor whose characteristic is shown in figure 5.4. This will generate a voltage ranging from about 0.4 V to 2.6 V over the temperature range 0-300°C. Little amplification is needed to provide a signal suitable for input to an ADC.

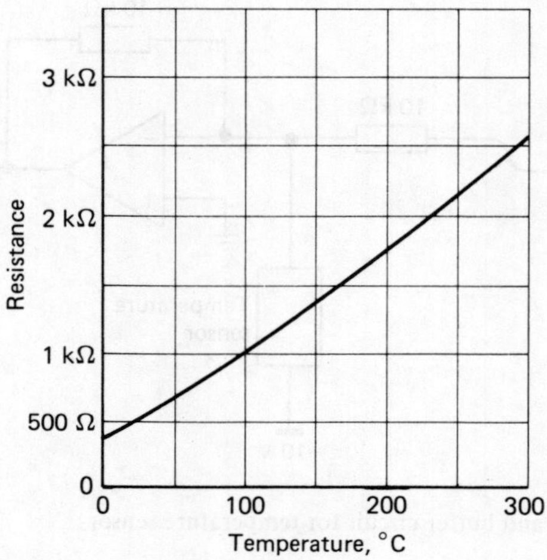

Figure 5.4 Resistance/temperature graph of silicon sensor

6

Force and Pressure Transducers

6.1 Strain gauges for force measurement

Many industrial microprocessor applications require the measurement of forces in structures or the weights of process materials etc. Forces are usually measured indirectly, one possibility being to measure the degree of stretching or compression they cause in connecting links that carry the forces. The basis of measurement is that the change in length is proportional to the force, up to a certain limit.

If the length of a link is L when unstressed, and it stretches by an amount δL under tension, the relative extension, or strain, is defined as

$$\epsilon = \frac{\delta L}{L}$$

This is generally specified in units of 10^{-6}, for convenience. Thus the elastic limit for mild steel used for construction work corresponds to a strain of about

$$\epsilon = 1250 \times 10^{-6}$$

For strains greater than this the material is permanently deformed and will not return to its original length when the tensile force is removed.

For any particular type of loading there is a relationship between strain and the stress in the material given by

$$\frac{\text{stress}}{\text{strain}} = E$$

E is a constant for a particular material, called Young's modulus. Thus if the strain ϵ is known, the stress is given by

$$\text{stress} = E \times \epsilon$$

Stress is defined as force per unit area, so that, finally, if the cross-sectional area of the link undergoing stress is A, we have

$$\text{stress} = \frac{F}{A}$$

Thus the force in the link is

$$F = A \times \text{stress} = A \times E \times \epsilon$$

Consequently if the strain can be determined, the force can be calculated.

Strain is usually measured electrically, using the fact that the resistance of a conductor changes when it is stretched or compressed. It is important to ensure that as far as possible strain is the only factor that causes a resistance change, so that the material used is chosen to have a low temperature coefficient of resistance – for example, copper-nickel or nickel-chromium alloys.

The conductor is usually fabricated as a thin foil bonded to a thin insulating film, which in turn is glued to the component under test. In order to minimise the effect of temperature, two foil strain gauge elements are often used differentially, one in compression and the other in tension. The layout of a typical foil gauge is shown in figure 6.1.

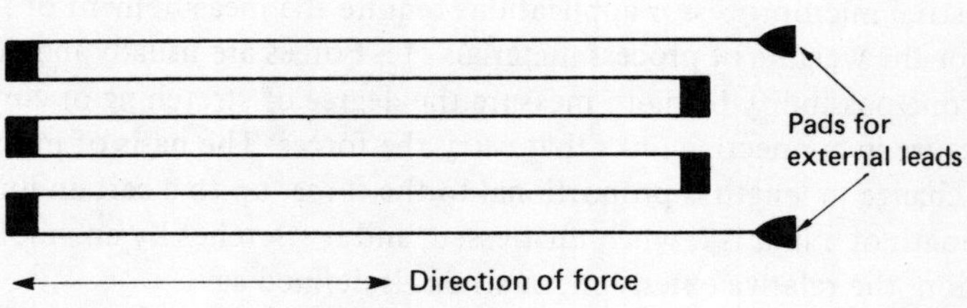

Figure 6.1 Conductor layout of foil strain gauge

For metal alloy strain gauges the relationship between resistance change and strain, called the gauge factor (G), is between 2 and 2.2. It is defined as the ratio of the relative change in resistance to the relative change in length. Thus

$$G = \frac{\delta R/R}{\delta L/L} = \frac{\delta R}{R \times \epsilon}$$

where the unstressed length and resistance are L and R and the changes under stress are δL and δR.

We can now finally relate the resistance change to the force in the test object since

$$F = A \times E \times \epsilon = \frac{A \times E}{G} \times \frac{\delta R}{R}$$

Although the force can be calculated using this expression, it is customary to check the calibration of the system by balancing the bridge circuit with load removed, then applying a known load and observing the unbalance voltage in the bridge circuit.

Where small values of strain are involved, higher bridge sensitivity is desirable. Semi-conductor strain gauges can then be used. These have gauge factors of typically 100, so they produce 50 times more unbalance signal than metal gauges for the same strain. Their disadvantage is their greater sensitivity to temperature changes.

Usually the steady-state component of force is of interest, so that a directly coupled amplifier with a small zero drift must be interposed between the strain gauge bridge and the ADC. However, for some purposes – for example, vibration studies – only the fluctuating component of the force is important. In such cases a.c. coupled amplifiers can be used, so avoiding errors arising from zero drift in the amplifier.

Strain gauge elements are normally made with nominal resistance values of 120 Ω, 350 Ω, 600 Ω and 1000 Ω. As with resistance thermometers, the current through them must be limited to 10–40 mA to avoid causing errors due to self-heating.

In addition to the resistance change in the gauge caused by temperature variation, a further error arises from differential expansion. If the temperature rises the gauge itself and the object to which it is attached would normally expand by different amounts, as their coefficients of expansion normally differ. However, when they are firmly glued together the gauge must expand by the same amount as the test object which is much larger in cross-section. This means that temperature changes alone will produce strains in the gauge and so an output from the bridge.

In order to avoid this error, manufacturers usually try to make the error caused by temperature strain balance the opposite error arising from the resistance change with temperature in an unstressed gauge. This balance will depend upon the material used for the test object, and manufacturers usually offer this automatic temperature compensation for three materials

mild steel
stainless steel
aluminium.

A further source of error arises from the change in resistance of the connecting leads with temperature changes. These are usually reduced by using additional compensating leads in another branch of the bridge circuit so that, so long as both sets of leads experience the same change in resistance, the bridge output will be unaffected. This condition can be satisfied by using identical wires for all of the leads, and twisting them tightly together so that they will have the same thermal environment.

In some cases the balancing can be carried further by using two gauges in opposition, one of which experiences tension and the other compression. This is possible where the load causes bending stress, as in a cantilever beam, or in a ring gauge.

The ring load cell is widely used for measuring compressive or tensile forces and enables four strain gauge elements to be built into a bridge circuit, so reducing considerably the effect of temperature changes. When the ring is subject to tension, it tends to elongate into an ellipse; the inner surfaces are stretched and the outer surfaces compressed. The four gauges are glued to the ring, as shown in figure 6.2, and wired into a bridge circuit, as shown in figure 6.3.

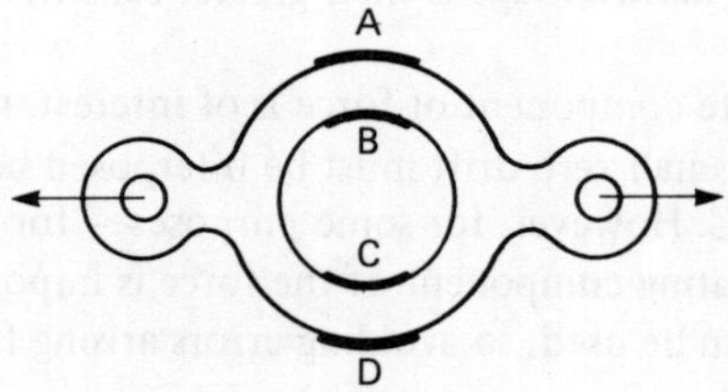

Figure 6.2 Attachment of strain gauges to ring load cell

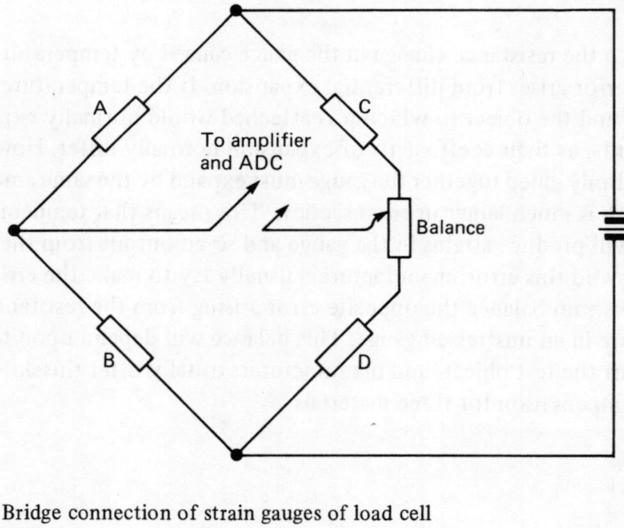

Figure 6.3 Bridge connection of strain gauges of load cell

A similar bridge circuit can be used for measuring the bending stress in a cantilever, with gauges A and D on one side of the beam and B and D attached to the other side.

Where tensile force is to be measured in, for example, a steel beam, the same bridge circuit can be used with gauges B and C attached to measure the tensile force, and gauges A and D attached perpendicularly to them. A and D will then measure the smaller compressive strain which occurs at right angles to the main tensile strain. The bridge connection will still largely balance out temperature effects.

6.2 Force balance weighing

Another technique that has been used for weighing is to suspend the load-carrying pan or platform on elastic supports. A sensitive position sensor, usually employing optical techniques, is arranged to detect any movement of the platform from its zero position. The output of the optical sensor is amplified and fed to a power amplifier which drives a moving coil actuator.

The actuator force opposes the load on the weighing platform, and with high amplifier gain in the feedback loop will be almost exactly equal to it. The actuator incorporates a stable permanent magnet, so that the force generated by the moving coil is proportional to the current I in the coil. By including a resistor of value R in series with the coil, the potential drop across it, IR, is proportional to the weight on the platform, and after amplification can be fed to an ADC for computer input. The direct system as described above is convenient for small loads but, for large loads, an arrangement of levers is used to reduce the force to a convenient level. The platform weight is then equal to the balance force multiplied by the mechanical advantage of the levers.

6.3 Pressure measurement

Strain gauges can be used for pressure measurement by measuring the strain in a thin diaphragm which sustains the pressure difference (figure 6.4). In this case the strain in the diaphragm is radial (assuming a circular diaphragm), so that special 'rosette' gauges are needed with the conductors disposed radially. Two gauges can be mounted on the diaphragm, one on either face, and connected as arms A and B in figure 6.3. Two fixed resistors take the place of gauges C and D. The arrangement of a rosette gauge is shown in figure 6.5.

An alternative arrangement is to use a more flexible capsule or bellows, which affords a greater movement under pressure, and attach a position transducer to it. The deflection will be proportional to pressure, and so in consequence will be the

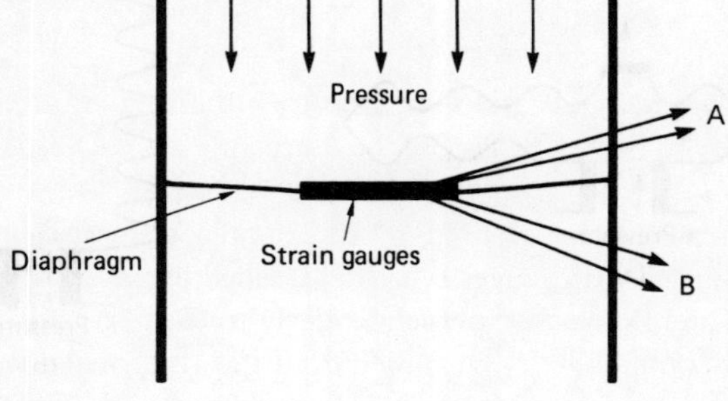

Figure 6.4 Diaphragm pressure gauge

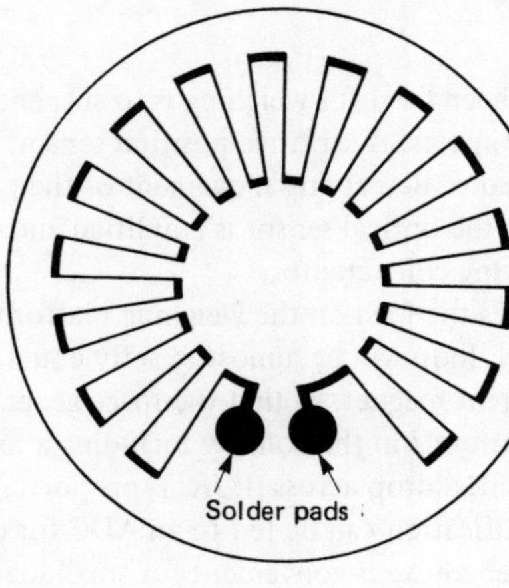

Figure 6.5 Layout of rosette gauge for measuring radial strain

output of the position transducer. Cross-sections of these two devices are shown in figure 6.6.

The LVDT is a convenient transducer for this purpose as it is linear, and can be made with a small stroke to match the expansion of the bellows.

Although the pressure transducers described above are adequate for a wide range of measurements, they have a limited frequency response on account of the mass of their moving parts.

Where rapidly varying pressure changes are under investigation, some alternative transducer is required. One possible device is the piezo-electric transducer. This consists of a thin plate of quartz or lead zirconate titanate ceramic which generates charges on opposite faces of the plate when stressed. The charge is very nearly proportional to the stress, and for a pressure transducer this is proportional to the differential pressure across the plate. Since the generated quantity is charge and

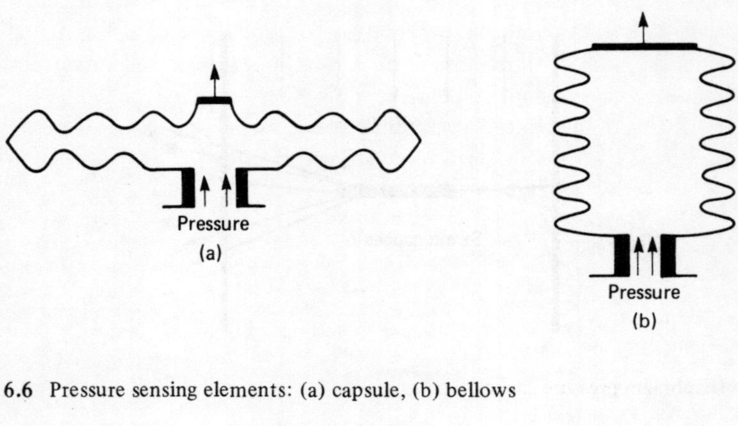

Figure 6.6 Pressure sensing elements: (a) capsule, (b) bellows

not current, a high input impedance amplifier is required, and a constant pressure yields no permanent output. The amplifier normally has a field effect transistor as the input stage.

The maximum frequency that can be handled depends upon the mechanical resonance of the transducer and its mounting, and can be up to 1 MHz. A typical application is to the measurement of pressure variations in the cylinder of a petrol or diesel engine.

For measuring much smaller pressure variations in gases, conventional capacitor microphones are often used. These can be built to respond to sub-audio frequencies down to less than 1 Hz – for example, for analysing shock waves caused by supersonic aircraft. At the other extreme, by reducing the diameter to 3 mm it is possible to raise their upper frequency limit to around 200 kHz.

6.4 Resonant beam sensors

For some process control applications it is necessary to measure the density of liquids or gases. Although this can be determined by weighing a known volume, an alternative technique is now available which is capable of continuous measurement in a stream of fluid flowing through the gauge. The method depends upon the mechanical resonance of a 'U'-shaped tube made of glass or light metal. Oscillation is maintained by an electronic amplifier and its frequency will depend upon the mass and stiffness of the tube. Although the stiffness is constant, the effective mass will change when the tube contains a fluid with a density different from that of air. By measuring the change in oscillation frequency when the fluid is admitted, the density of the fluid in the tube can be deduced. The effect of introducing a liquid will obviously be greater than that caused by a gas, but since frequency is the physical quantity that can be measured with the greatest precision by comparatively simple electronic counting circuits the method has been widely adopted. For liquids, an instrument having a relative density range of 0.5 to 2.0 is quoted as having an accuracy of better than 1 per cent.

The calibration depends upon temperature, so that in some instruments a microprocessor is used to compute temperature from readings of a thermistor sensor. The temperature of the sample is displayed, and this is used to correct the frequency reading and to calculate density.

An interface to a data collecting microprocessor is available either in BCD parallel form from the LCD display, or in serial form via an RS232 connection.

7

Flow Measurement

7.1 Digital flowmeters

Flow measurement is usually concerned with the volume of fluid passing through a pipe or channel per second. In some cases this can be measured directly, but it is generally easier to measure the velocity of the fluid at some point and deduce from this the volume flow rate.

Direct measurement can be performed by directing the fluid through cylinders which contain pistons restricted to a fixed stroke. Volume of flow is then proportional to the number of piston strokes. To smooth out the flow two cylinders are often used, one filling while the other exhausts. The major objection to these devices is that they can handle only a limited flow rate, and they involve a significant pressure drop.

Flow rate can be measured indirectly by measuring fluid velocity, for example, by suspending in the fluid a small turbine or rotor which has propellor blades attached to it. A magnetic pickoff can be provided by embodying a small permanent magnet in the rotor and mounting a reed relay next to the tube wall. The relay will close briefly during each revolution of the rotor. This arrangement requires a non-magnetic tube – for example, of copper, brass or some plastic material – and since the magnet can be embedded inside the rotor no extra obstacle to fluid flow is caused by the rotary pickoff. Flow rate can be measured by counting the number of relay closures that occur in a known time. This is a very convenient arrangement for microprocessor input, since only one digital input is needed. One line of an input port could be used or, if other actions were also required, one section of a counter–timer package could be left to count pulses, perhaps using another section to generate a fixed time interval. The only main program action required would be to start the counter and the timer.

If the end of the time interval is programmed to create an interrupt, the interrupt service program need only read the counter to determine the flow rate, store this for later processing, reset the timer, enable interrupts and return to the main program.

Other pickoff arrangements have been used, for example, a coil outside the tube will have an e.m.f. induced in it each time the magnet passes near it, or a Hall effect sensor can be fixed to the outside of the tube.

To improve accuracy, some flow-straightening vanes may be mounted upstream of the rotor to remove any rotary motion of the fluid which could introduce errors. Also the calibration of these turbine transducers is accurate only if the fluid is flowing smoothly without eddies or turbulence (so-called 'laminar flow').

In some industrial operations where small but controlled amounts of liquid are added to a process, a metering pump is often used with a variable speed motor drive. This type of pump is designed to deliver a fixed volume of liquid per stroke over a wide range of speeds. In order to measure either the flow rate from the pump or the total volume added over a period, it is usually sufficient to count the number of strokes of the pump, either by fitting a sensor to the pump or to the motor that drives it. This could be a magnetic or optical pickoff that generates one pulse per stroke. The flow rate is then proportional to the pulse frequency, and the total volume delivered is proportional to the total pulse count over a period.

In order to simplify interfacing, digital flowmeters are generally preferred for microprocessor input. However, some convenient types of flow transducer deliver an analogue output, and these will require some signal amplification followed by an ADC.

7.2 Pressure sensing flowmeters

A number of practically important flow or velocity transducers measure fluid velocity by sensing pressure or pressure difference. A simple arrangement is the vane type of anemometer used mainly for measuring the velocity of gases. This comprises a small flat plate suspended perpendicularly to the gas flow by a spring-mounted arm. The pressure on the plate, and hence the arm deflection, depend upon the gas velocity. A displacement transducer can be fitted to the arm to measure deflection. By using a stiff spring and a very light arm in conjunction with a sensitive inductive pickoff, it is possible to follow pressure fluctuations with a frequency of 100 Hz. The turbine type of flowmeter cannot follow any such rapid fluctuations and is better suited to almost steady-state measurement.

Where the nature of the fluid could cause problems with an obstruction in the flow, like a vane, other means can be used which do not involve measuring devices suspended in the fluid. By introducing some constriction in the tube carrying the fluid, the velocity of the fluid will be increased, since for a constant flow rate the mass of fluid flowing across all cross-sections of the tube must be the same. In the case of a liquid which is almost incompressible, the volume flow per second must remain constant. Thus the product of the cross-sectional area of the pipe, A, and the average velocity of the liquid at that section, v, must be constant. Consequently if the cross-section is reduced by a factor of 2 by a tapering section of pipe, at the narrow section the mean velocity will be doubled. This will alter the pressure in the pipe.

In order to determine the pressure change, we use a theorem first proposed by Bernoulli and named after him. This states that in a fluid flowing smoothly in a frictionless pipe the total energy of any volume of fluid is constant.

The energy comprises three components

Kinetic energy
Pressure energy
Potential energy.

If we assume that the pipe is horizontal, or the two sections are close together, the potential energy due to height above some datum level can be considered constant. The equation then reduces to the statement

kinetic energy + pressure energy = constant

or

$$\tfrac{1}{2}\rho V^2 \quad + \quad P \quad = \text{constant}$$

If the pipe is arranged to taper down and then return to its original section, as shown in figure 7.1, constituting a venturi, the pressures and velocities at the up-stream point and at the throat, or narrowest point, will be connected by the equation

$$\tfrac{1}{2}\rho V_1^2 + P_1 = \tfrac{1}{2}\rho V_2^2 + P_2 = \text{constant}$$

If the area of the throat is K times the pipe area ($K < 1$) then

$$A_2 = KA_1$$

$$V_2 = \frac{V_1}{K}$$

Thus

$$\tfrac{1}{2}\rho V_1^2 + \rho_1 = \tfrac{1}{2}\frac{\rho V_1^2}{K^2} + \rho_2$$

whence

$$P_1 - P_2 = \tfrac{1}{2}\rho V_1^2\left(\frac{1}{K^2} - 1\right)$$

Here A, V and P are the cross-sectional area, the mean velocity and the pressure of the fluid, respectively. Suffix 1 relates to the up-stream conditions, suffix 2 to conditions at the throat.

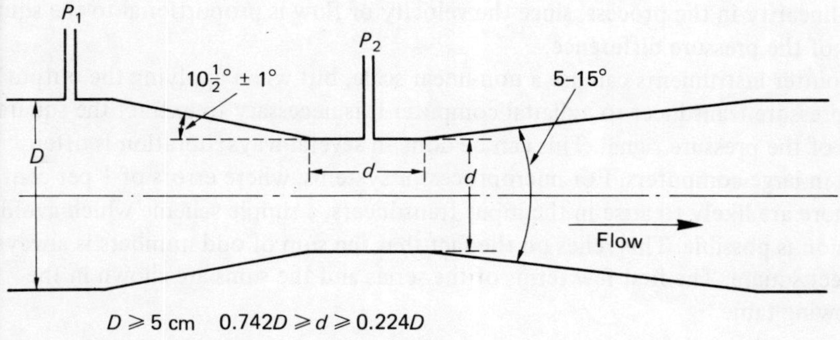

Figure 7.1 Cross-section of venturi to BS 1042

By measuring the differential pressure $\Delta P = P_1 - P_2$, the fluid velocity can be deduced since all other quantities in the equation are known. Thus

$$\Delta P = \tfrac{1}{2}\rho V_1^2 \left(\frac{1}{K^2} - 1\right)$$

or

$$V_1 = C\,\frac{2 \times \Delta P}{\sqrt{\rho}}$$

where C is a constant equal to $K/\sqrt{(1 - K^2)}$.

This calculation is of course based upon ideal conditions. In practice the velocity will be slightly lower than that calculated, because there is some energy loss due to turbulence, roughness of the tube walls, etc.

Although the venturi tube is the preferred arrangement for flow measurement, since it causes least obstruction and pressure drop, it is necessarily rather long and, where space is important rather than minimum energy loss, other throttling methods can be used which involve introducing some obstruction into the parallel-sided tube. One arrangement is the orifice plate, which is a circular plate usually bolted between the flanges that couple two sections of pipe together. The plate has an orifice smaller than the cross-section of the pipe, and the pressure drop across the plate is the measured quantity. A similar arrangement which creates rather less turbulence is the nozzle. This is a short tapered section of tube which is inserted in the main pipe.

The preferred sizes and shapes of orifice plates, nozzles and venturi tubes are given in BS 1042 which deals generally with methods of measuring fluid flow in closed conduits. The calculation of fluid velocity from pressure drop given above is correct only for incompressible fluids. It can thus be used for all normal liquids, but may need some correction when used for gases.

One difficulty with the venturi method of flow measurement is the essential non-linearity in the process, since the velocity of flow is proportional to the square root of the pressure difference.

Pointer instruments can use a non-linear scale, but when applying the output of the pressure transducer to a digital computer it is necessary to extract the square root of the pressure signal. This can be done in several ways; iteration is often used in large computers. For microprocessor systems, where errors of 1 per cent or more are likely to arise in the input transducers, a simple scheme which avoids division is possible. This relies on the fact that the sum of odd numbers is always a perfect square. The first few terms of the series and the sums are shown in the following table.

Terms	1	3	5	7	9	11	13	15
Sums	1	4	9	16	25	36	49	64

Algebraically, the sum of n terms of an arithmetic series is given by

$$S_n = \frac{n}{2}(2a + (n-1)d)$$

where a = first term
d = common difference.

Here $$a = 1, d = 2,$$

so $$S_n = \frac{n}{2}(2 + (n-1)2)$$

$$= n^2$$

Thus a simple computer program can be written which computes each sum in turn, and compares it with the number representing pressure. As soon as a sum is discovered that is greater than the pressure, the operation stops and the required square root is equal to the number of terms. To give adequate accuracy in the result it is best to use double-length (16-bit) numbers for the sum, scaling the pressure suitably.

7.3 Ultrasonic flow measurement

Ultrasonic pressure waves in liquids can be used to measure their velocity. One technique is to time the passage of a pulse of ultrasonic energy along a path nearly parallel to the direction of flow.

If the spacing between the transmitting and receiving ultrasonic transducers is L, and the velocity of the pressure wave in the liquid is C, the time taken for the pulse to travel from transmitter to receiver is

$$t = \frac{L}{C}$$

However, if the liquid is moving with velocity v in the same direction as the pulse of ultrasound, the velocity of the pulse with respect to the tube containing it is $C + V$, and the transit time will be

$$t_1 = \frac{L}{C + V}$$

The simplest way to measure t and t_1 is by the 'sing-around' technique, by which the pulse is made self-sustaining. As each pulse is received it is amplified, shortened to a rectangular waveform, and used to trigger another pulse from the transmitter. Thus, neglecting the signal delay in the electronic circuits, the frequency of the pulses will be

$$f_1 = \frac{C + V}{L}$$

Knowing C and L, this expression could be used to calculate V, but this would be rather inaccurate as the relative change in frequency is quite small, and the velocity of sound C is rather temperature dependent.

However, by using another beam transmitted in the opposite direction, over the same path length, the frequency will be

$$f_2 = \frac{C - V}{L}$$

If we now mix these two frequencies together we obtain a beat note at the difference frequency, $f' = 2V/L$. This does not directly involve the temperature-dependent propagation velocity C, and so will be a more reliable measurement. It is also one that can easily be made using a counter–timer peripheral package, using one channel for counting and a second channel to generate a fixed time interval.

Another method using ultrasonic sensing is the vortex flowmeter. This relies on the fact that if a small obstruction, usually in the form of a thin post, is placed in a smoothly flowing fluid, a series of vortices will be formed downstream of the obstruction. The important feature is that for given conditions the distance along the fluid between successive vortices remains constant for a wide range of fluid velocities. Thus the fluid velocity is proportional to the number of vortices passing any point in the tube per second.

The vortices are sensed by sending an ultrasonic beam across the tube containing the fluid. When a vortex crosses the beam it scatters some of the sound energy and so the strength of the received signal falls. These dips in the signal can be amplified, squared up, and sent to a counter–timer peripheral package. The number of pulses counted in a given time is then proportional to the fluid velocity.

For convenience, the post, transducers and electronic system can all be packaged together and attached to a short section of pipe with standard flanges. This extra section of pipe can then be inserted into the existing pipe run with a minimum of disturbance.

7.4 Thermal flow transducers

For specialised flow measurements where very rapid response is required, thermal transducers can be used. These comprise either very thin wires or thin films of metal deposited on a ceramic substrate. The transducers are made from a metal or alloy with a large temperature coefficient of resistance, and heated up by passing a fixed current through them. The equilibrium temperature they attain depends upon the velocity of the fluid passing over them. Hence by measuring their resistance the flow rate can be deduced, once the transducer has been calibrated.

These transducers have been used to measure the very rapid pressure fluctuations in the neighbourhood of turbine blades. One method of easing the read-out is to build the transducer into an *RC* Wien bridge used to determine the oscillation frequency of a power amplifier. Changes in flow rate will then alter the temperature of the transducer, hence its resistance, and so the oscillation frequency of the amplifier. Thus the flow measurement is reduced to a measurement of oscillator frequency; this is a digital quantity and simple to determine using a counter–timer package. The only other requirement is to stabilise the output amplitude of the amplifiers, and thus the power dissipated in the transducer.

7.5 Use of correlation techniques

Although the flow measuring methods described in previous sections of this chapter can deal with nearly all cases of gas or liquid flow, they are either ineffective or very inaccurate for liquids that contain a high proportion of suspended solids. Unfortunately a number of industrial processes require just such a measurement, hitherto rarely tackled for lack of a suitable method.

The availability of inexpensive microprocessors has now enabled these awkward measurements to be made much more accurately, a particular example being the measurement of the flow rate of raw sewage by optical sensing. The proportion and types of suspended solids prevent any use of suspended rotor or ultrasonic sensing methods. The procedure used measures the amount of light reflected by a narrow strip of the liquid surface whose length is perpendicular to the flow direction. The result is a widely fluctuating and irregular signal. However, the material is flowing smoothly, and a similar optical system, placed a few feet downstream, will pick up a very similar signal, but delayed by the time taken by the liquid to flow between the two test sites. The arrangement is shown in figure 7.2.

If now we could take a short section of the first waveform and delay it by the transport time between the test sites, it would match very well the waveform then emerging from the second site. This fact can be used to determine the transport

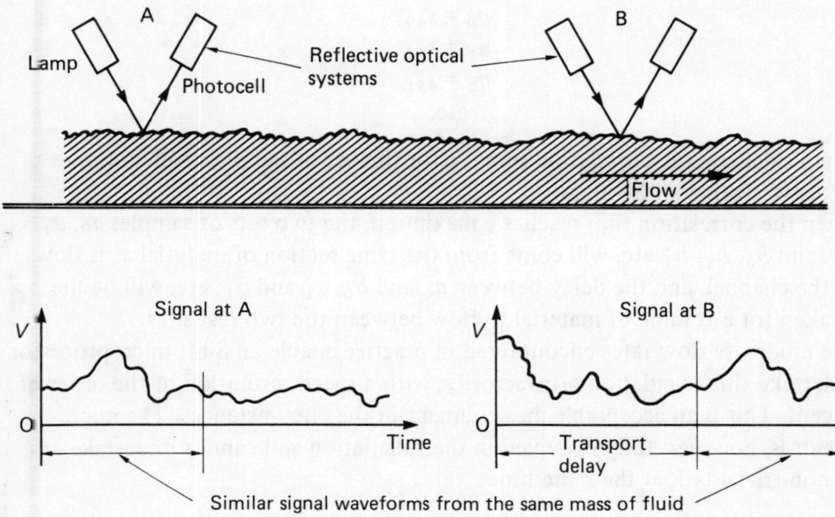

Figure 7.2 Correlation measuring of flow

time, and thus the flow rate, using 20–30 samples of the optical sensor waveform. These are taken at both sites and compared. It is necessary to store, say, 30 samples of the second waveform, but more of the first waveform. The comparison between the samples is done by a correlation process.

If the second set of samples is b_0, b_1, b_2, b_3, b_4, etc., we select an equally long run of successive samples of the first waveform and evaluate the correlation sum as

$$\Sigma = a_0 b_0 + a_1 b_1 + a_2 b_2 \text{ etc.}$$

where the first set of samples is

$$a_0, a_1, a_2, a_3 \text{ etc.}$$

The correlation sum is evaluated for a range of the a_0, a_1 samples, starting with a set a little earlier than the set b_0, b_1, b_2, etc. and increasing the time delay between a_0 and b_0 each time.

For example, if we take, say, 100 samples of the first waveform, and label them

$$s_0, s_1, s_2, s_3 - s_{99}$$

we may take the first set of samples as

$$\begin{aligned} a_0 &= s_{50} \\ a_1 &= s_{51} \\ a_2 &= s_{52} \\ &\text{etc.} \end{aligned}$$

Having calculated one value of the correlation sum, say Σ_0, we take the next range of samples as

$$a_0 = s_{49}$$
$$a_1 = s_{50}$$
$$a_2 = s_{51}$$
etc.

and then calculate the correlation sum Σ_1 for this set of samples, and continue the process, storing all Σ values.

When the correlation sum reaches a maximum, the two sets of samples a_0, a_1, a_2 etc. and b_0, b_1, b_2 etc. will come from the same section of material as it flows along the channel, and the delay between a_0 and b_0, a_1 and b_1, etc. will be the time taken for a volume of material to flow between the two test sites.

The moderate flow rates encountered in practice enable an 8-bit microprocessor to undertake this calculation satisfactorily, with a speed resolution of the order of 3 per cent. This is an acceptable measurement in the circumstances. The microprocessor is, however, fully occupied in the calculation and cannot undertake any other non-trivial task at the same time.

8

Velocity and Acceleration Measurement

8.1 Measuring rotational speed

Many control systems require the measurement of the speed of a shaft or other rotating component. Where this information is an input to a microprocessor system a digital measurement will be preferred since it avoids handling analogue signals and the expense of an ADC. The usual technique is to generate a train of pulses whose frequency is proportional to angular speed. This means generating a fixed number of pulses per revolution. In a slowly changing system where adequate time can be devoted to the measurement, it may be possible to generate only one pulse per revolution. This leads to a simple transducer comprising a fixed reed relay and a small magnet mounted on the shaft. Once per revolution the magnet sweeps past the relay closely enough to close the normally open contacts.

Alternatively an optical system could be used, consisting of a light emitting diode mounted near to a phototransistor. Between the two a disc, fixed to the shaft, rotates. Cutting a radial slit in the disc will produce one pulse per revolution. These pulses can be counted for a fixed interval of time to determine speed.

One objection to this scheme is the time taken to make an accurate measurement. For example, to determine angular speed with an error not exceeding 1 per cent we must count at least 100 revolutions of the shaft. When this measurement is part of a speed control system, the time taken may be excessive and lead to an unstable system. If the motor is rotating at 1000 r.p.m., this measurement will take 100/1000 minutes = 6 s. Thus the control system can make a correction to the motor drive power only every 6 s, by which time the speed error may be excessive. Unless the system inertia is very large, this delay in assessing speed and using the error as a feedback term will result in an unstable system, and some faster method is needed.

One possible scheme would be to use a fast clock signal of known frequency, feeding a high speed counter to measure accurately the time of one revolution. This provides a rapid determination of speed, but involves additional hardware and a parallel interface to handle the output of the counter. A much more economical method of reducing the measurement time is to generate many more pulses per revolution by increasing the number of slits in the optical disc. Thus if we have an optical system with 100 slits in the disc, we shall generate 100 pulses per revolution and we can determine the speed with the same accuracy in only 1 revolution, that is, in 60 ms, with a shaft speed of 1000 r.p.m. By using the microprocessor and its program to count the pulses, we avoid the need for extra hardware, and need only one input line to handle the data.

In this example we have assumed a fairly coarse measurement of speed with an error around 1 per cent. Should greater accuracy be required, even more pulses per revolution may be needed. For example, the optical discs used for controlling the capstan speed in precision tape recorders are produced photographically and may have up to 1000 black bars on a transparent background to permit fast and accurate speed measurement.

Where such accuracy is not required, and perhaps 50–100 pulses per revolution are adequate, a magnetic pickup can be used. This consists of a thin pole piece on which a coil of wire is wound. The pole tip is fixed near to the teeth of a steel gear wheel so that, as each tooth moves past it, a small pulse is induced in the coil. Some magnetising force is needed; this can be supplied either by a permanent magnet, or by passing a small current through the coil. Although the magnetic pickoff does not have the high resolution possible in an optical system, it will work satisfactorily when covered with grease, dust and any non-magnetic debris. It may thus be more reliable in a hostile factory environment than an optical system which must be sealed against dust and dirt.

A magnetic pickoff having 180 teeth is often used in car engines that have electronic ignition. This gives a pulse every 2° of crankshaft rotation, and the pulse train can be used to give a fairly rapid indication of engine speed. For timing the ignition it is necessary to relate to crank or piston position. Thus an additional pickoff is provided which indicates the top dead centre position of the crankshaft. The 2° pulses then serve another function, since by counting them after a crankshaft pulse the crankshaft position at any time can be determined.

Where some transducer has to be attached to an existing system, it is often not possible to add components such as a slotted disc or a magnetic pickoff. One possibility then is to paint a set of black and white stripes on a shaft or a shaft coupling and use a reflective optical system. This involves projecting a small beam of light on the stripes and picking up the light reflected from them. For all of these optical systems there is some advantage in using infra-red light. It is then easy to mask out visible light, so ensuring that ambient light changes, fluorescent lights etc. will not generate spurious signals.

Where a slotted disc or an optical disc can be attached to the shaft, a convenient packaged optical sensor is available. This consists of a light emitting diode and a phototransistor packaged in a 'U'-shaped moulding. The diode lies in one limb and the phototransistor in the other, and the disc periphery rotates between them. The plastic used is opaque to visible light but transparent to the infra-red radiation emitted by the LED. The sensor thus operates reliably regardless of changes in ambient light, but can be overloaded by direct sunlight which contains a large proportion of infra-red energy. The sensor is nevertheless much more convenient to use than one based on visible light which must be shielded from stray illumination.

8.2 Tachometer generators

In some control applications it is necessary to extract what data is available from existing transducers and measurement systems. Many variable speed electric motor drives incorporate tachometer generators to indicate speed. These are small permanent magnet alternators whose output voltage is almost exactly proportional to speed. In order to mask noise from slip rings, the full speed output is usually of the order of 120 V. Thus in order to use this for microprocessor input it must be rectified and smoothed, and reduced to 5 V or 10 V by a potential divider. It can then be connected to an ADC. Alternatively, since the frequency of the output is equally proportional to speed, a proportion of the output signal can be fed to a trigger circuit which generates a fixed amplitude rectangular wave and then to a counter–timer package to measure frequency. In order to improve the resolution a trigger circuit can be connected to each phase of the generator output (usually a three-phase signal).

When using tachometer voltage as an indication of speed, there is a degree of conflict between the rapidity of response and the resolution obtainable. For example, if it is necessary to determine the speed to within $\frac{1}{2}$ per cent, it is also necessary to attach sufficient smoothing to the rectifier circuit to reduce the ripple signal to less than $\frac{1}{2}$ per cent. Inevitably this means that the smoothed signal cannot respond very rapidly to speed changes. The situation is eased by using three-phase rectification which reduces the ripple voltage and increases its frequency.

A much more rapid measurement of speed can be made by generating timing signals for the zero crossings of the a.c. output of the tachometer, and measuring the period with a high frequency timing supply. For example, if the tachometer output frequency is 50 Hz, the time between zero crossings is 10 ms. Using a clock frequency of 100 kHz, 1000 clock pulses will be counted between successive zero crossings. Allowing an error of ±1 count, this gives a speed resolution of about ±0.1 per cent.

On account of their cost and the driving power they require, tachometer generators are generally used only with fairly large motor drives. For small motors some other scheme is preferred.

8.3 Using motor e.m.f. to measure speed

In a d.c. shunt-wound motor, often used for variable speed drives, the e.m.f. generated by the armature is proportional to the product of angular speed and field flux. Thus if the field flux is held constant, the armature e.m.f. should be proportional to speed. Consequently if this e.m.f. can be measured, the speed can be determined without adding any additional transducer. Unfortunately the only voltage readily measurable is V, the terminal voltage supplied to the armature. The e.m.f. can then be determined as

$$E = V - I_a R_a$$

where I_a and R_a are the armature current and resistance. Thus knowing R_a, a measurement of I_a will allow the voltage drop $I_a R_a$ to be calculated. One technique is to insert a small resistor in series with the armature, and measure alternately the armature voltage V and the IR drop. A simple calculation in the microprocessor will enable the e.m.f. to be determined. This scheme is simplified if the resistor inserted has a value equal to R_a, when only subtraction is needed. However, the armature resistance R_a is liable to increase by 20 per cent or more as armature temperature rises, and for accurate measurement this temperature dependence must be incorporated into the measurement circuit.

For small d.c. motors, such as those used in chopper circuits for variable speed operation, another technique has been used. These change the speed by changing the ratio of on-time to off-time in the armature circuit, which is fed from a constant voltage source via the chopper. With this arrangement the armature e.m.f. can be measured directly by taking a short sample of the armature voltage when the chopper is not conducting, and thus there is no armature current flowing. Consequently there is no IR voltage drop. This scheme has limited application because armature inductance delays the decay of the armature current and the measurement of e.m.f. is not accurate until the current has decayed to nearly zero. It can, however, be used with small motors having relatively low inductance.

The use of generated e.m.f. as an indication of the speed of a d.c. motor with fixed excitation is adequate for many industrial drives, but it suffers from the fact that, even with constant field current, the field flux is not quite constant. This is because the armature current, which depends upon the load torque, also modifies the flux in the machine air gap. This effect is called 'armature reaction', and it causes a small drop in air-gap flux as the armature current increases. The effect of armature reaction can be reduced, if $I_a R_a$ compensation is used, by a slight adjustment to the magnitude of the $I_a R_a$ term. Complete cancellation of the armature reaction effect is not however possible, since it is a non-linear function of armature current, and the $I_a R_a$ compensation term is a linear one.

8.4 Linear velocity measurement

Although most velocity measurements involve rotary motion, a few applications require linear velocity measurement. One of these arises in vibration measurement. This involves the use of a scaled-up moving coil loudspeaker movement, normally having a permanent magnetic field. A small search coil wound over or alongside the main motor coil can be used to measure linear velocity, which is proportional to the e.m.f. induced in the coil. This is, of course, an analogue output. For a sine wave excitation it is usually convenient to use a peak detector and feed this value to an ADC which in turn feeds the microprocessor. However, for some tests, a random waveform may be used in order to excite all likely resonances, and then it is necessary to take a number of samples at short intervals of time in order to record accurately the velocity as a function of time.

The same velocity measuring technique can be used in vibration studies where the vibration is generated by mechanical means. By using a heavy permanent magnet, or adding mass to it, we produce a mass large enough not to move significantly. The moving coil is suspended with a fairly high compliance, and is coupled to the vibrating test object through a light stiff rod. The moving coil thus moves in sympathy with the test object and the inertia of the magnet ensures that it remains still. Thus the e.m.f. induced in the coil is, as before, proportional to the linear velocity of the coil. The general arrangement is shown in figure 8.1.

If a position transducer is fitted it is possible to calculate velocity either by analogue differentiation of the analogue position signal or, if a series of digital measurements of position have been taken, numerical differentiation can be performed by the microprocessor program. The basic difficulty with these schemes is that any noise or irregularity in the position signal is accentuated by the differentiation. This is evident in a potentiometer position transducer, where the noise arising from the sliding contact is emphasised when computing velocity. For this reason a transducer that gives directly a signal proportional to velocity is generally preferred.

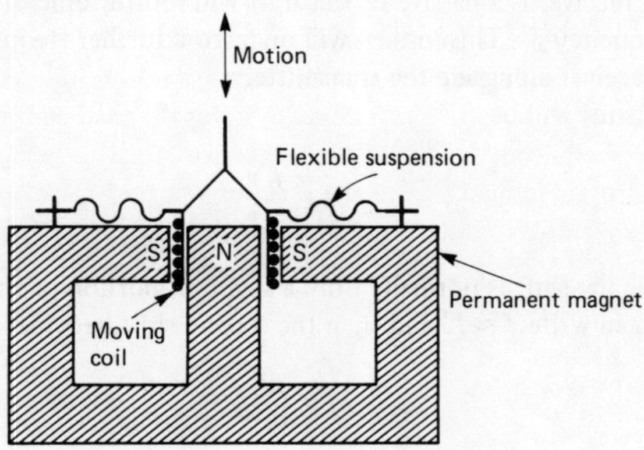

Figure 8.1 Cross-section of moving coil velocity transducer

8.5 Doppler speed measurement

The so-called 'Doppler shift' phenomenon is used for measuring speed in some applications. It relies on the fact that when the transmitter of a periodic signal in a medium is moving with respect to the receiver, the received frequency differs from that observed when the transmitter is stationary. The effect is particularly marked when high speed aircraft fly past. As they pass the listener, the pitch of the engine noise falls very noticeably. To determine speed, the effect must be quantified. If we assume that the transmitter is at rest, and the receiver is moving towards it with a velocity V, we can determine the apparent frequency at the receiver by counting the number of cycles of the periodic signal that pass the detector in 1 second.

If the receiver were stationary, in one second it would detect $f = c/\lambda$ cycles where c is the velocity of the wave and λ the wavelength. If the receiver is approaching the transmitter with velocity v, it will move a distance V in a second. This corresponds to v/λ wavelengths. Thus the receiver will detect an additional v/λ cycles due to its motion, giving a total number of cycles detected in a second as

$$\frac{v}{\lambda} + \frac{c}{\lambda}$$

The frequency observed by the receiver is thus

$$f' = \frac{c + v}{\lambda}$$

The frequency shift due to the motion is thus

$$f' - f = \frac{c + v}{\lambda} - \frac{c}{\lambda} = \frac{v}{\lambda} = \frac{fv}{c}$$

A similar shift is observed if the transmitter is moving towards a stationary receiver. If the receiver is a passive reflector, it will return some of the incident energy at a frequency f'. This in turn will undergo a further frequency shift when it arrives at a receiver alongside the transmitter.

The second shift will be

$$\delta f = \frac{f'v}{c}$$

Since the frequency shift is normally only a small proportion of the emitted frequency, we can write $f \approx f'$ and then the second shift becomes

$$\delta f = \frac{fv}{c}$$

The total shift is then twice this, or

$$\frac{2fv}{c}$$

As an example, for a microwave Doppler sensor operating at 10 GHz, a car moving at 60 m.p.h. would generate a Doppler shift of about 1790 Hz.

Usually a proportion of the transmitted signal is mixed with the received signal and filtered. The major component of the output is then the Doppler frequency. If this is taken to a trigger circuit to convert it to a rectangular wave, we have a signal which can be fed directly to a counter-timer package. This can readily measure its frequency and so determine speed.

The same microwave system can be used to measure the velocity of material flowing along a channel, provided that it has enough surface irregularity to reflect some of the incident signal.

Doppler speed measurement can also be used with other types of wave propagation, for example, arterial blood velocity has been measured using an ultrasonic transducer, and the velocity of solids or liquids can be measured using a laser light source.

8.6 Measuring acceleration

In the design of mechanical systems it is often necessary to measure acceleration. This is usually done by measuring force, and relying on the relationship between force and acceleration

$$F = ma$$

where a is the acceleration of a mass m, and F is the impressed force required.

The measurement of acceleration is then reduced to the measurement of force. If the mass is suspended by a compliant beam, strain gauges can be attached to the beam to determine force. However, a widely used arrangement which avoids the need for a compliant mounting and reduces the possibility of resonance effects uses piezo-electric crystals for measuring force. These are stiff, so that the resonant frequency of the transducer is raised to typically 20–30 kHz. This is far above the frequencies normally encountered in vibration studies, and so no error will be introduced by resonance effects.

The piezo-electric material is usually operated in the shear mode, to reduce the unwanted signals caused by temperature transients. One design uses a central post which is part of the mounting base of the transducer. The piezo-electric element is ring shaped, bonded to the central post and also to an outside ring which comprises the mass, as shown in figure 8.2.

An alternative design which avoids the need for adhesive is the triangular configuration of three plates, each with its own mass, in which an outside ring holds the piezo-electric plates in firm contact with the central post and the three masses.

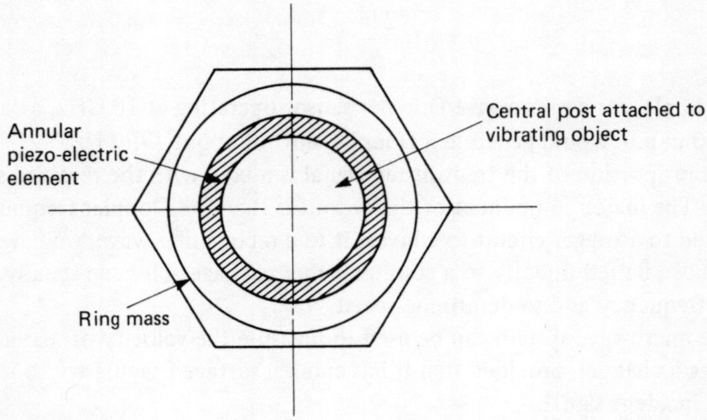

Figure 8.2 Accelerometer using piezo-electric element in shear mode

This is shown in figure 8.3. The initial stress in the outer ring pre-loads the plates and is claimed to improve the linearity.

A typical transducer of this type has a charge sensitivity of about 20 pC for an acceleration of *g*. With a typical capacitance, including the output cable, of about 1000 pF, the corresponding voltage output would be about 20 mV. The sensitivity can be increased by using a larger mass, but this will reduce the resonant frequency.

This measuring technique is of necessity an analogue process, so that the output must be taken through an ADC before it can be fed into a parallel I/O package.

If transducers have been fitted to measure velocity, acceleration can be calculated by a process of numerical differentiation. As with the calculation of speed from position, this process tends to accentuate the components of noise already present in the system. A particular case of this is in the control of large passenger lifts

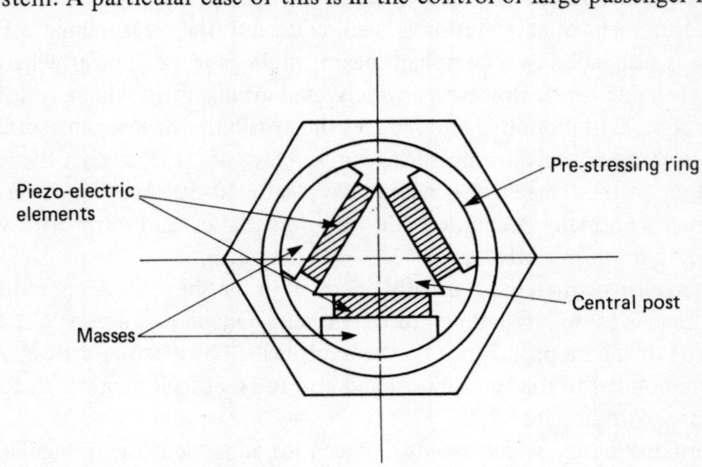

Figure 8.3 Triangular accelerometer with pre-stressed elements

where it is important to limit the acceleration and thus the disturbance to passengers. Since tachometers are usually fitted as part of the motor speed control system, the speed signal can be differentiated to determine acceleration.

9

Proximity Sensing

9.1 Applications of proximity sensing

Earlier chapters of this book dealing with position sensing have dealt only with the continuous measurement of the position of an object within a given range of movement. There are many applications for some much simpler 'go–no go' measurement. For these the output is a single bit of information which indicates that the object has reached or passed a particular position. Where adequate power is available a mechanical switch may be used, operated by pressure from the moving object, as in the 'over-ride' switches often fitted to lifts as a safety measure. Similar arrangements may be used in which a swinging arm operates a micro-switch for sensing that the paper in a printer has run out, or the end of the tape has been reached in a tape recorder.

However, there are many other applications in which the mechanical load imposed in operating a switch would be unacceptable, and some form of non-contact sensing is required. This is the function of proximity sensors which can use many of the techniques embodied in position sensing. The requirements are much less critical than those needed for continuous position sensing, for example, there is no call for linearity, but the threshold position at which the output transition from 0 to 1 occurs may need to be accurately defined and stable. In order to avoid rapid fluctuations in the output when the system stops in the threshold position, it is usual to provide some hysteresis in the output electrically. This means that, for example, if the output changes from 0 to 1 at a spacing of 20 mm, as the object approaches the sensor it must move back to a spacing greater than 20 mm before the output reverts to 0.

Some of the devices mentioned in chapter 4 for measuring angular speed are in effect proximity sensors, for example, the use of a small magnet attached to a rotating shaft or a turbine rotor suspended in a tube carrying liquid. When the magnet moves near to a reed relay the contacts will close and indicate the proximity. Equally the interruption of a beam of radiation between an LED and a photodiode by a vane attached to the moving object can be used to indicate proximity. Capacitance sensing has also been used where the moving object is either a conductor or can be coated with a thin metal foil.

9.2 Inductive and capacitative proximity sensors

Inductive proximity sensors are often used since they impose no load or friction on the object sensed, and will work regardless of dust, dirt, oil, water or most other materials encountered in a hostile environment. The only essential is that the object sensed must either be a conductor, or have a conducting plate attached to it.

These devices incorporate a tuned oscillator with the resonant coil being mainly air-cored and mounted next to the insulated outer surface of the sensor. The oscillator has enough feedback just to sustain oscillation. When a conducting surface approaches the sensor, eddy currents are induced in it and the resultant damping of the tuned circuit reduces the amplitude of the oscillation sharply. This signal change is amplified to provide the output.

With a mild steel object, the sensing distance for a 1 mm thick target varies between 2 and 15 mm for various types of sensor. With non-ferrous metals, the sensing distance is reduced by a factor of 2 to 3.

An alternative technique for proximity sensing is to use capacitance effects. These are usually based upon a balanced a.c. bridge, one of the capacitors in the bridge being arranged to have an external stray field. When any material that is a conductor, or a dielectric having a permittivity greater than that of air, is moved near to the capacitor, the bridge will be unbalanced. The out-of-balance signal can be amplified and fed to a trigger circuit to provide the output. Capacitative sensors are more expensive than inductive sensors on account of their greater complexity, but they can be operated by a wider range of materials. These include powders, glass, wood, PVC etc., in addition to ferrous and non-ferrous metals. They can also be made to have somewhat larger sensing distances than inductive sensors.

9.3 The Hall effect and its applications

When electrons moving through a vacuum pass through a magnetic field perpendicular to their direction of motion, they experience a force which deflects them in a plane perpendicular to both their initial motion and to the magnetic field. This is the basis of the deflection coils that are used to move the luminescent spot so as to produce the picture in a television receiver. The same force acts upon an electron moving in a metal but, as it collides frequently with the atoms of the metal, the deflection observed is very much smaller. The phenomenon is called the Hall effect, after the scientist who discovered it.

The effect may be increased by a factor of many thousands by using a semi-conductor material such as indium antimonide, but currently many Hall effect devices use silicon as the semi-conductor material so that amplifiers, trigger circuits etc. can be integrated on the same chip. The Hall effect voltage is smaller with silicon but is generally in the range of millivolts.

The effect is usually sensed by passing a current through a thin film of material which has electrodes attached on either side, opposite one another as shown in figure 9.1. If these are located accurately, there will be no potential difference between them due to the current flow, which is perpendicular to the line joining the contacts. However, if a magnetic field is introduced perpendicular to the plane of the film, the lines of flow of the current will be curved, as shown in figure 9.2, and there will be a potential difference between the electrodes. This is roughly proportional to the product of the magnetic field strength and the current.

In the Hall effect integrated sensor, a voltage regulator on the chip provides a constant current for the sensor, and a stabilised supply for the voltage amplifier which handles the output of the voltage probes. The amplifier feeds a Schmitt trigger circuit with an open collector output stage. A typical device has hysteresis and will operate with a flux density of about 0.05 T (500 gauss) and release with about half of this.

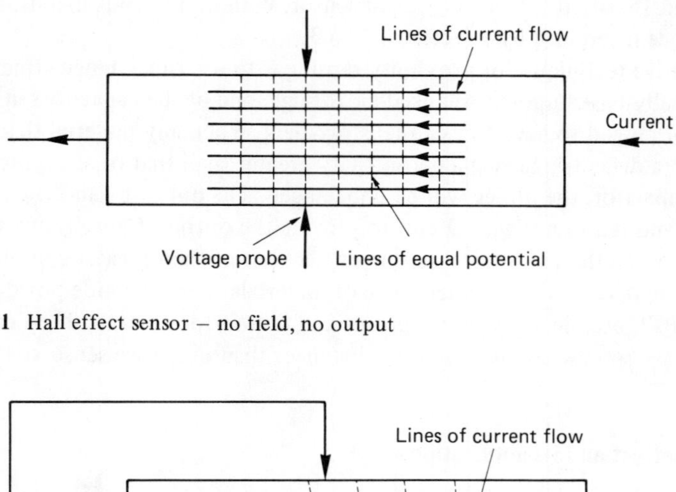

Figure 9.1 Hall effect sensor – no field, no output

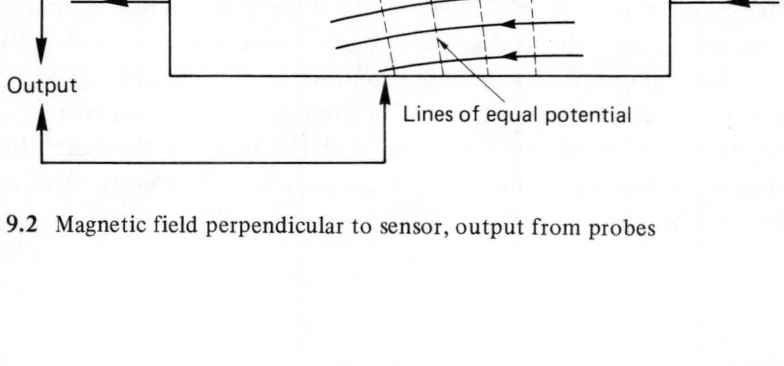

Figure 9.2 Magnetic field perpendicular to sensor, output from probes

The advantages of Hall effect sensors of this type are high speed – they can respond at up to 100 kHz – and the lack of bouncing. Thus when connected to electronic circuits or computer inputs, there is no need to include 'de-bouncing' arrangements into either hardware or software. Since they do not involve any moving parts their reliability is very high, and they have a major area of application in explosive atmospheres where mechanical switches which may produce arcs cannot be used. They are used in the keys of computer terminals and electric typewriters, and for many proximity sensing applications. Their use has increased now that fully integrated devices, built on a single chip, are available, as they cost no more than a reed relay.

9.4 Touch pads

The introduction of CMOS logic packages, which have extremely high input resistance, enables circuit designers to produce circuits that need very small input currents. Such circuits can be influenced by a touch from the finger tip. In most environments the stray 50 Hz fields from nearby mains wiring will inject a 50 Hz signal into any high impedance point that is touched. This can be amplified and rectified to produce a logic signal. Alternatively a d.c. shift to some point on the circuit may be caused by connecting this point to some potential via the finger tip. This involves providing two metal strips close together which are bridged by the finger tip.

One such arrangement is shown in figure 9.3. Here the gate input is held high by the 33 MΩ resistor. When the key pad is touched the finger tip connects the input point to earth, with a resistance in the region of 100 kΩ to 1 MΩ. This will pull the gate input below the logic threshold, so that the output of the inverter will be a positive pulse.

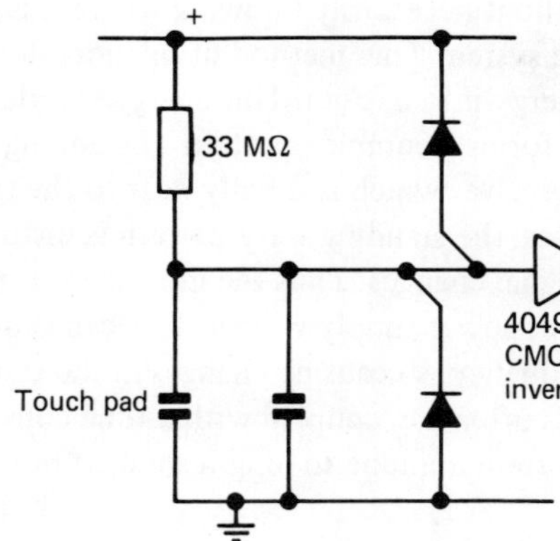

Figure 9.3 Circuit for touch pad input

If the computer program is waiting for the touch input it will normally be testing the inverter at frequent intervals and will respond within, say, a millisecond of touching the pad.

On the other hand, if we are concerned with setting up conditions that the microprocessor will read in later, when some other activity has ended, it may be necessary to store the data, and the touch pad may be used to set the state of a bistable which will be cleared subsequently by the microprocessor. This can be arranged by using two cross-coupled NAND gates for the bistable, as shown in figure 9.4.

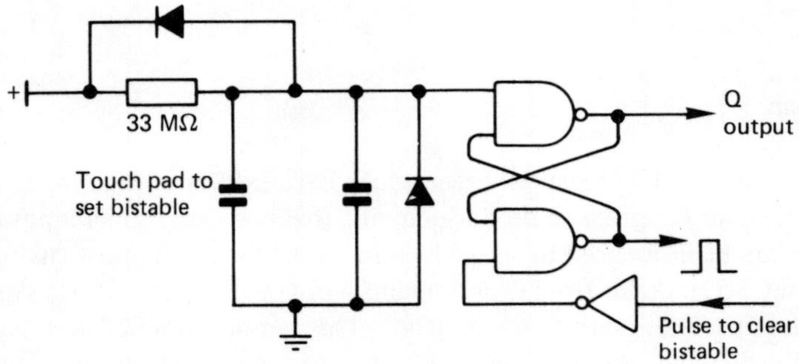

Figure 9.4 Circuit for setting bistable by touch pad

9.5 Standing wave sensors

Limit switches and proximity sensors are used to detect when an object has reached a specific position. There is also a requirement to detect when any object has entered a defined space, without necessarily knowing where it is. This is the basic need in any intruder sensing system. One method of intruder detection involves radiating a wide beam of energy into a room. This energy is reflected from the walls and other objects and forms a complex pattern of standing waves. Some of the energy is returned to a receiver which is usually near to the transmitter. When any object in the room moves, the standing wave pattern is disturbed, and the amplitude of the received signal changes. Thus the movement can be detected. Slow changes in temperature, power supply voltage, etc. can cause corresponding changes in the oscillator frequency, so causing changes in the standing wave pattern. Thus the detector is often a.c. coupled with a time constant of some minutes. This renders the system immune to long-term drift but sensitive to any intruder.

The radiation may be provided either by a microwave source such as a Gunn diode operated at about 10 GHz, or an ultrasonic transducer. Both of these produce energy with wavelengths of 3-10 cm, thus giving good sensitivity to small movements.

The microwave sources typically produce about 10 mW of r.f. output, and contain a receiver cavity into which is coupled a small fraction of the transmitted signal. A mixer diode combines this with the received signal to give an output signal at the Doppler shift frequency. Thus these sources are normally used with an a.f. amplifier and detector whereby any output denotes some movement in the protected area. This technique has the advantage that slow changes in oscillator frequency will not give any output, and will thus not result in false alarms.

Many burglar alarm systems now include microprocessors which combine inputs from a number of different transducers and can make a response based upon the nature of the input signals. They may also include automatic operation of warning signals, closure of doors, and the transmission of a message by telephone to the police.

10

Data Collection Systems

10.1 Optical sensing

In many computer systems there is a need to gather data from a plant, factory or store. We have discussed the collection of small volumes of data from analogue inputs, sensing switches, etc., but in some applications the input may involve a block of data. For this, special input devices have been developed, some of which include writable data stores. One example which uses input material that can be generated without any special instrument or machinery involves optical sensing. This uses a document which has a number of rectangular spaces for the user to enter data. The spaces are either left blank or filled in with a pencil mark. The form is subsequently scanned in a machine to detect whether or not each space has been filled in, and the results are stored in a computer. These forms can be used for stores records, questionnaires, multiple choice examination papers etc.

Similar optical coding has been used where information has to be collected from moving objects. For example, reflecting strips attached to railway wagons have been used to provide information about the type of wagon and its destination, to allow trains to be assembled automatically in microprocessor-controlled railway marshalling yards. The wagons are released one at a time down a slight slope and, as the data from each one is fed to the microprocessor, the program decides where it should go and operates points accordingly. Other position sensors follow the motion of the wagon until it reaches the correct siding.

Optical sensing has also been used in a scheme for controlling buses. For a limited run, bus position can be estimated by counting wheel revolutions, but inevitably errors accumulate owing to manoeuvring in traffic and the elasticity of the tyres, and it is necessary to establish position periodically by gathering data from optical systems mounted on the pavement. A microprocessor installed in the bus collects and interprets the data, and is connected to a radio transmitter and receiver. This enables a central controller to interrogate each bus to request its position or to transmit a message which is displayed to the driver. Each bus has its own identification code which precedes all messages. This is set up on switches in the bus and the microprocessor is programmed to ignore all messages that are not preceded by the identification code. Each time the bus position is established by the optical sensors, the counter that totals wheel revolution is reset to zero, and the dead reckoning of position starts from the new datum point.

A further application of optical sensing is found in the 'light pen' often used in conjunction with cathode ray tube (CRT) displays. This consists of an optical system and a small phototransistor, arranged to pick up light from the CRT raster when held close to it. A spring-loaded switch is usually incorporated so that the operator can signal to the microprocessor when the light pen is correctly positioned on the screen. By observing which character is being displayed when the light pen pulse is received, the microprocessor program can determine where the pen has been placed, and so what action is required. The pen can be used, for example, to select one option from a menu of possible actions displayed on the screen. Also, if supported by appropriate graphics software, the light pen can be used to 'draw' diagrams and illustrations on the display. In some versions of light pen, the phototransistor and its associated amplifier are built into the display cabinet, and coupled to the light pen through a metre of optical fibre or 'light pipe'.

10.2 Serial data input

Where data must be entered into a microprocessor system manually, this can be done conveniently via a keyboard. For solely numerical data a 10-way key pad will suffice but, where text input is involved, a conventional 'QWERTY' keyboard is generally used. Clearly it would be very costly to interface this to a computer on a wire per key basis as this may need over 50 lines, which would require four parallel interface packages. Since the keyboard is often remote from the microprocessor it is much cheaper to use a single pair of wires and transmit data serially. Techniques and signal codes for this were developed many years ago for electromechanical devices such as teleprinters. On account of the mechanical limitations the data rates were restricted to the range 45–110 bits per s. When computers were developed, and users looked for a convenient input device, the teleprinter was the obvious choice, since it includes an output printer. Although still used, teleprinters have largely given way to visual display units comprising a keyboard and CRT display. These are smaller, lighter, cheaper and, being entirely electronic, they can handle much faster data rates, up to 2400 or 4800 bits per s. However, they provide no printed record of the output, so where this is required a printer must be connected to the microprocessor. The electro-mechanical printers had only mechanical governors to control their speed, so they used an 'asynchronous' mode of operation. In this the timing of any character is not related to the timing of adjacent characters, so synchronism between transmitter and receiver need be maintained only for one character. The character code generally used is the American ASCII code, now standardised internationally as ISO-7 code. It consists of 7 data bits, with provision for an 8th parity or check bit, preceded by a start bit and followed by two stop bits. Thus each character comprises in all 11 bits. Such a code matches very well with 8-bit microprocessors, and all manufacturers provide an interface package suitable for serial data communication. These packages are programmable and allow the program to specify whether 5, 6, 7 or 8 data bits are used, if a parity bit is used, and if so whether odd or even parity. (This means that the data and parity bits together contain either an odd or an even number of ones.)

At the receiving end of the line, the interface package strips off the start and stop bits, checks the number of data digits and the parity, if used, and assembles the data as a parallel input byte. Usually the package is programmed to create an interrupt when the byte is available, and the interrupt service routine will read the byte, store it away, and return control to the main program. The interface package also has a transmit side which accepts a parallel byte, adds start and stop bits and a parity bit, if required, and then sends the assembled character out serially. The transmit and receiving sides of the package can operate independently to give full duplex working.

In many applications the transmit side is programmed to create an interrupt whenever it has begun sending a character out; this indicates that the transmit buffer is empty and the interrupt service routine can extract the next character from the microprocessor and load it into the serial data package. In addition to the normal connections to the data and address buses, the package needs a clock supply. This can be either at the transmitted bit rate or 16 times higher. Packages that perform this serial/parallel data conversion include the 6850 (for the 6800 family), the 8251A (for the 8080/8085 family) and the Z8440 (for the Z-80).

Where a teleprinter is connected to the packages directly, and not via a telephone circuit, some signal level conversion is generally required. The output of the communications packages is a TTL level signal whereas the teleprinter may require either a current loop drive or a V24 (RS232) circuit.

The 20 mA current loop circuit was used for connecting teleprinters to early minicomputers, having the advantage that the teleprinter is a purely passive device – the supply for both transmit and receive circuits comes from the computer. This asymmetry between the two sides of the circuit is awkward for testing since the computer output cannot be looped to the computer input to test the program and interface. Equally the computer cannot be switched to 'Local Copy', when the printer echoes each key pressed, without an extra power supply. Also the comparatively high current of 20 mA means that the resistance of the line between teleprinter and microprocessor must be fairly low, hence the distance between them must be short.

The V24 circuit standard is an international one for connecting together data originating and data transmission equipments. The signal levels are nominally ±12 V with the receiver resistance lying between 4 kΩ and 7 kΩ. Consequently the currents are small, and there is also a requirement for current limiting in that the maximum short-circuit current flowing in a short-circuit between any two terminals must not exceed 25 mA. Many teleprinters now adopt this standard, as do nearly all visual display units (VDUs). It is also the international standard for feeding data signals into the modulators and demodulators (modems) needed for sending data over a telephone line.

Clearly some level conversion is needed between TTL levels and V24 signals. Special integrated circuits are now available for this, for example, the MC1488 and MC1489A, and the 20 mA current drive for a teleprinter can be provided by optical couplers such as the 4N33.

The smaller currents involved in V24 operation and the relatively large signal amplitude mean that the lengths of the lines to the terminal may be increased to 300 metres or more.

10.3 Synchronous operation

The asynchronous mode of operation described above is essential when electro-mechanical devices such as teleprinters are used, and is also generally adopted for purely electronic terminals such as VDUs for data rates up to 600 or 1200 bits per second.

As transmission rates increase, circuit costs tend also to increase, and there is a greater need for their efficient use. The asynchronous mode can be criticised in that, even at full speed, out of every 11 bits only 8 at most carry data. Efficiency would be increased if the start and stop bits could be eliminated.

This would be possible if a continuous stream of bits were transmitted, and at the receiving terminal a clock supply could be incorporated which could extract a clock pulse train, synchronous with the received data. By using a quartz crystal to control the transmit bit rate a very stable clock supply can be generated, and it is then possible to lock a local oscillator on to this frequency at the receiver using a voltage-controlled oscillator and a phase lock loop.

A major problem with this scheme is that we need some mechanism to determine where one character ends and the next one starts. This is provided by programming the receiving communications package to enter the 'search mode' initially. In this the received bit stream is fed into a shift register (usually 16 bits wide) and logic examines this register continuously to determine when its contents match two 8-bit SYN characters previously loaded into the package by the initialising program.

As soon as these characters are detected a counter in the package is cleared, and every 8th clock pulse it subsequently accepts will denote that a complete character has arrived and is ready for transfer to the microprocessor. The detection of the SYN characters normally switches the receiving package into the 'data transfer' mode. During the search phase the bit stream being received is not sent on to the microprocessor.

To ensure reliable synchronisation, the originating terminal normally sends four SYN characters and it is also designed to insert further SYN characters should there be a gap in the stream of incoming data. This is necessary as the system works only if a continuous bit stream is transmitted.

There is no general agreement among manufacturers about the provision of both synchronous and asynchronous communication. Some microprocessor families provide a single package which can be programmed for either mode, for example, the 8251A for the 8080/8085 series, and the Z8440 for the Z-80. On the other hand two separate packages, the 6850 (asynchronous) and the 6852 (synchronous) are provided for the 6800 family.

10.4 Bar code readers

A form of optical data input which is being increasingly used is the bar code reader. This uses a code comprising black and white bars printed on some object. The code is read either manually, by passing an optical wand over the code pattern, or by a fixed reader below which objects are moved by a conveyor belt. Typical applications are in point-of-sale terminals where each item sold carries its identity on a bar code. The code is read by the microprocessor in the terminal and the price of the item is obtained from a look-up table in the microprocessor store. To allow for price changes, etc. the table is loaded each morning from a central computer.

Another application is to the control of books in a library. Each book has a bar code strip fixed inside the cover, and each user has a plastic card which carries a bar code pattern identifying him. By scanning both the book and the plastic card a borrower can be associated with the books he borrows, reminders can be printed when the books have been out for a prescribed time, and any fines due can be calculated. In addition the time taken to operate the system is much reduced, and so is the space needed for data storage.

The optical wand is about the size of a ballpoint pen, and comprises a lamp and a phototransistor together with their optical systems. Lenses are needed to focus the light into a small spot since the bar width may be no more than 0.25 mm. A lens is also needed to gather as much as possible of the reflected light and focus it on to the phototransistor. In order to ensure that the bar code is always at the correct distance from the lenses, the wand is designed to be used with the tip in contact with the paper.

The software for decoding the bar code is fairly complex since the speed at which the pattern is scanned can vary quite widely, and may in fact change as the code is being read.

A typical bar code consists of an initial guard pattern of 101, five 7-bit characters, a central guard band of 01010, a second group of 5 characters, a check character and the right-hand guard pattern 101. Each character consists of 7 bits which are chosen to produce two black bands (bit value = 1) and two white bands (bit value = 0). Each band may comprise 1, 2, 3 or 4 bits. Figure 10.1 shows the coding for the decimal digits 1 and 3.

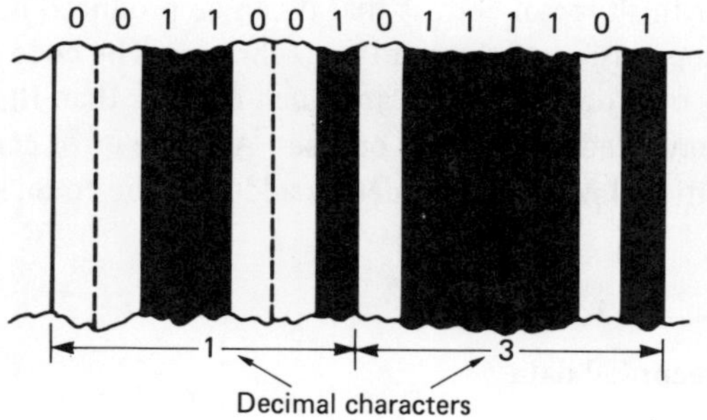

Figure 10.1 Example of 7-bit bar code

One decoding procedure involves the use of a short timing loop to determine the time at which each signal transition occurs. Since the first guard band generates three transitions (0 → 1, 1 → 0, and 0 → 1) and each of the next 5 characters scanned generates 4 transitions, we can estimate the time per bit by dividing the time between the first and 23rd transitions by 38, the number of bits in the guard band, and the first 5 characters. This information can then be used to decode each of the characters.

In order to decode the signals from the wand it is necessary to determine when the transitions occur in each character. This can be done by first determining the average bit length. In the first 5 characters, each character is coded so that it always begins with a light band and ends with a dark band. There is thus a 1 → 0 transition at the beginning and end of every character, and a second 1 → 0 transition somewhere within the character. Since the initial guard band carries the 3-bit code 101, and is preceded by a white area, the 2nd, 4th, 6th etc. 1 → 0 transitions will indicate the start of a character.

The decoding program can include a short timing loop which is run continuously and operates a counter. Each signal transition causes the current value of the counter to be stored for subsequent processing. The average duration of a bit can then be calculated by dividing the time between the first and the 23rd transitions by 38.

Since each character starts with at least one 0, we can now reconstruct each character from a knowledge of the moments when transitions occur. All bits are 0 up to the first transition, and are then 1 until the second transition, etc. Having reconstructed the 7-bit format for each character, a look-up table enables the program to determine the corresponding character. A similar procedure can be used for the second 5 characters, except that the central guard band is coded 01010, and an additional check character is added after the 5 data characters.

When the 10 decimal digits contained in the code have been determined, the price of the item can be found by examining a further table within the microprocessor store.

The advantage of this type of code is that it can be produced in a range of sizes, depending upon the precision of the printing technique. The code itself can be printed by a variety of processes, which are much cheaper than those needed to make magnetically encoded cards, tapes or discs. Also the error checking process will detect any additional marks subsequently added to the code, so preventing fraud.

10.5 Magnetically recorded data

Bar code reading is a reliable process if the medium carrying the code is clean and not folded. This is usually the case for point-of-sale applications, or the code fixed inside library books. However when documents may be handled, creased or folded, other methods which do not rely on reflected light are preferred. The main such method uses a magnetic code, read by magnetising the material and then passing it over a replay head similar to that used in a tape recorder. In one application the characters are printed in magnetic ink, which includes a very finely divided iron powder. A special character set was developed for this which can be read both visually and magnetically. The major use of this process lies in encoding the branch account and bank information on the bottom of each bank cheque. These can then be sorted automatically.

Printing data in magnetic ink has the major advantage that existing machinery can be used to apply the information to the document. However, once printed the data cannot be altered. Some applications involve modifying data each time the item carrying the code is used. For this a continuous strip of magnetic tape can be attached to a plastic card. An example of this is the plastic card used for telephone calls. These are bought for a certain sum which is encoded on the card. Each time the card is used the sum available is reduced by the cost of the call made, and the remaining sum can be displayed to the owner. It is also possible to record details of each call on the card, so that when the sum originally encoded has been expended, a print-out of the calls made can be provided. This application requires a microprocessor in the telephone to handle the reading and writing of the data and the charge calculations, and to control the use of the telephone.

A similar plastic card is used in bank cashpoints, but in a read-only mode. The details of the holder's bank, account number and brach are encoded on the card, together with a personal identity code. To avoid fraud and misuse, this identity code must be keyed into the cash point before any transaction is allowed.

Probably the simplest use of magnetic encoding is that provided on tickets used on the London Underground System. Here a digital code is recorded on the back of the ticket, which is read by the machine controlling the entrance gate and then erased to prevent the ticket being used again.

11

Output Transducers

11.1 Power requirements

The nature of the output facilities provided for microprocessors can be divided into two broad categories, those that provide information, and those that provide action. Generally the provision of information is a low power activity, and it is possible to connect directly to the microprocessor output packages. This would be the case, for example, when sending analogue voltage signals to a pointer meter, a chart recorder, or a CRT display via a DAC.

Where the microprocessor is used to control motors, heating elements, large contactors, etc. at power levels in the kilowatt range, a large power gain is needed. The microprocessor and its associated packages all operate at low power levels and can typically drive one or two standard TTL gates. This represents a power level in the region of 10 mW (2 mA at 5 V). Thus to control a 10 kW motor a power gain of 10^6 is needed. Clearly, for an output transducer which requires substantial power a major feature of the interface design will be the provision of adequate power gain. The details of the circuit depend upon the power level required, and the nature of the signal (analogue or digital). For moderate powers of tens of watts, relatively cheap transistors can be used for both the linear and digital outputs. In general the power dissipation of the transistor must be much greater for an analogue output than for a digital signal.

As an example of an analogue output, we take the driving of a solenoid-operated valve. The valve is spring-loaded, and the amount of valve opening is roughly proportional to the current fed to the solenoid. If this has a full load rating of 1 A at 24 V, and is connected in series with a transistor across a 24 V supply, the maximum power dissipation in the transistor will occur when the 24 V supply is shared equally by the solenoid and the transistor.

In this case the solenoid resistance is

$$R = \frac{24}{1} = 24\ \Omega$$

The maximum current is 1 A, and the current for maximum transistor dissipation is half of this. Thus the peak transistor dissipation is

$$P = \tfrac{1}{2} \times 12 = 6 \text{ watts}$$

Since the transistor may have to maintain this current and power level for an indefinite time, its continuous rating must be 6 watts.

In contrast, should the solenoid be used to actuate a contactor, it will handle only digital signals, and will be either fully energised from a 24 V supply or unenergised. When operated, the current flowing will be very nearly 1 A, and the voltage across the transistor will be its saturation voltage, typically in the range 0.7–1 V. Thus the maximum transistor dissipation will not exceed 1 watt (I_C = 1 A, V_{CE} = 1 V), compared with 6 watts when used for analogue output at the same maximum power level of 24 watts output. The advantage of the digital output becomes greater as the supply voltage increases.

Although the duty imposed upon the transistor is clearly less onerous with digital output, since the peak transistor dissipation is much lower, the current rating of the transistor (here 1 A peak) is the same in both cases. A smaller heat sink will, however, be required for the digital output case.

The above comparison between the transistor dissipation with analogue and digital signals assumes that a single transistor is used to drive the load. Where a large current gain is required, two transistors in a Darlington connection can be used. These are made in a single package, and have a total current gain in the region of 500–1000. They are the obvious choice where a large current gain is needed, but for switching mode operation they have a slight disadvantage. Owing to the circuit used, the saturation voltage is in the range 2–3 V, so that their power dissipation is several times larger than that of a single transistor used with the same load current.

The above discussion of transistor power dissipation is concerned only with the steady-state conditions. When the load current is switched on and off repeatedly at a high frequency, extra transistor dissipation is caused during the transitions from cut-off to saturation and in the reverse direction. This extra dissipation must be considered in the thermal design of the output stage, and arises from the finite time required to turn the transistor current on or off. During this time the voltage across the transistor is changing from the supply voltage to the saturation voltage, and taking the simplest case of a purely resistive load the transistor dissipation will be greatest at the mid-point of the transition when both collector–emitter voltage and collector current are half their maximum values.

The voltage, current and power variations with time are shown in figure 11.1. If we make the further simplifying assumption that the collector current rises and falls at a constant rate, as depicted in figure 11.1, it can be shown that the average power dissipated during a transition is

$$P = \frac{1}{6} P_{max}$$

where P_{max} is the dissipation of the load when fully energised. Thus if the transition time is t_r and the switching frequency is f, the additional power dissipated in the transistor will be

$$P_s = \frac{1}{3} f \times t_r \times P_{max}$$

since one cycle includes both a rising and a falling current transition.

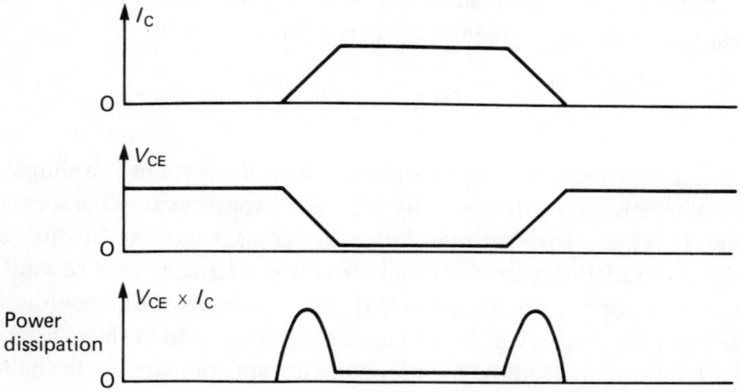

Figure 11.1 Current, voltage and power waveforms when switching a resistive load

This dissipation increases both with the transition time and with the switching frequency. The transition time with a resistive load depends mainly upon the transistor characteristics, but this situation is found only when driving LEDs and filament lamps. In the great majority of cases the transistor load will be an inductive device such as a relay or a contactor. Then the characteristics of the load circuit, particularly the ratio of inductance to resistance, will largely determine the duration of the transition.

In all cases where the transistor load is inductive a protection diode should be connected across it, as shown in figure 11.2, to avoid generating excessive voltages which could cause transistor failure when the transistor current is cut off.

In practical situations the extra transistor dissipation caused by switching can usually be ignored when driving electro-mechanical devices such as relays, contactors, solenoids etc., since the frequencies involved are never more than a few Hz. However, when high switching speeds, which may reach tens of kHz, are involved – for example, in high frequency inverters – the extra loss becomes important.

11.2 Switched mode operation

When an analogue output is required, at power levels above some tens of watts, a conventional class A output stage causes difficulties on account of its poor efficiency and the consequent need to design for high transistor dissipation. The transistor losses can be reduced considerably by using a 'chopper' output stage with pulse width control.

This involves switching the output transistor on and off rapidly, and controlling the ratio of on-time to off-time. The switching rate is usually ten or more times greater than the highest frequency present in the output signal. Assuming a steady-state condition in which the on-time is t_1 s and the off-time t_2 s, during the complete cycle time of $t_1 + t_2$ the load current will flow for a time t_1. If the load current is I_0 during this time, the mean current flowing will be

$$I = I_0 \times \frac{t_1}{t_1 + t_2}$$

If $t_1 + t_2$ is kept constant, this is proportional to t_1. Thus we have a simple proportional control which can be driven easily from a microprocessor. One scheme, which allows for about $\frac{1}{2}$ per cent resolution in the mean current, involves an 8-bit counter which is initially cleared, and the zero output signal is used to send a logical 1 to an output pin, which controls the drive transistor. The counter is incremented at a fixed rate, which can be set by a delay loop in the program.

After each incrementation the counter contents are compared with the required output value (assumed to be also an 8-bit number) and when equality is detected the output to the drive transistor is changed to logic 0, so cutting off the transistor current. The counter continues incrementing and, when it turns over from 255_{10} to 00_{10}, the drive current is once more turned on.

Thus if the number representing the required analogue output is N, the mean output current will be

$$I = I_0 \times \frac{N}{256}$$

where I_0 is the current that flows when the transistor is turned on.

If the microprocessor has other tasks, it would be better to use a counter–timer for the generation of the delay, and create an interrupt when output operations are required. Some counter-timer packages are designed for pulse width modulation, and can be programmed to generate the required signal for the drive transistor. After the initial program sequence, the only data required is the number N. Once this has been loaded into the package it will continue to generate the width-modulated pulse train with a constant mean value which will not alter until a new value of N is entered.

The advantage of switched mode operation is that the transistor dissipation is much reduced. However, extra smoothing may be required to reduce the output ripple to an acceptable level. The amount of smoothing needed is smaller as the switching frequency is increased, but this requires faster transistors and may generate an unwanted r.f. noise.

11.3 Triac and thyristor control

Many applications involving power control require either a variable a.c. supply or a variable d.c. supply.

Variable d.c. supplies for the field or armature circuits of d.c. machines can conveniently be provided using thyristors as controlled rectifiers. These are switching devices which are normally open-circuit, but turn on rapidly when the gate electrode is energised. This requires a pulse of 30–100 mA at several volts, but to ensure reliable triggering a train of pulses is often used. Given an anode circuit that can sustain a current of this order, the anode–cathode path switches to low impedance, with a voltage drop of 1.5–2 V, and remains in this state until the anode current falls to some tens of mA when the thyristor reverts to the high impedance state.

Each thyristor can be used as a rectifier; it conducts only when the anode is positive to the cathode, and a trigger pulse has been applied, and ceases to conduct at the end of the half-cycle when the supply voltage falls near to zero. By sending the trigger pulse at the beginning of the half-cycle, full conduction and maximum output can be obtained. As the firing pulse is delayed with respect to the zero voltage point at the beginning of the half-cycle, the mean and the r.m.s. load current will fall. Thus to control the load current or power we again need a variable delay, but now full output corresponds to zero delay, and zero output to a delay of one half-cycle, or 10 ms for a 50 Hz supply.

Although driving signals for the gate of the thyristor can be derived from a counter–timer package, the arrangements differ slightly from those needed for pulse width modulation of a d.c. supply since the starting point for the delay is now an external event, the zero crossing point of the a.c. supply waveform. Since most supply authorities control the mains frequency to close limits, a fixed period can be used as the increment of time. For example, if an 8-bit number is used to specify the thyristor output, the timing increment for a maximum of a half-cycle delay at 50 Hz is

$$\frac{10 \text{ ms}}{256} \simeq 39 \ \mu\text{s}$$

corresponding to a frequency of about 25.6 kHz.

Where a counter–timer package is used to generate the delay, an oscillator providing this frequency is needed, otherwise if the delay is generated by software, a counting subroutine that takes 39 μs can be used, and called the requisite number of times.

The maximum output is obtained when the delay is minimum (say, 1 unit), half power is available when the delay is $\frac{1}{4}$ cycle, or 128_{10} (80 hex) units, and minimum output when the delay is maximum, corresponding to 255_{10} or FF hex units.

Thus if the computer program controlling the system has generated an error signal E which represents the power output required, E must first be scaled so that it lies in the range $0–255_{10}$, then subtracted from 256_{10}. This gives a value which denotes the number of 39 μs intervals of delay needed between the beginning of the half-cycle and the firing pulse. This can be generated by the program, and a firing pulse produced by taking one output line high for, say, 20 μs, then returning it to the zero state. This could be used to drive a transistor into conduction. Alternatively, a simpler connection with more power gain can be provided by interposing an open collector inverter between the output package and the thyristor drive transistor. For more reliable triggering, a small program loop which generates, say, 20 pulses would be preferable.

Using the inverter requires the opposite polarity of output from the peripheral package, so that the output is normally high, and is taken low for 20 μs and then returned to high to generate an output pulse, as shown in figure 11.4.

Where the mains frequency is liable to appreciable variation, the scheme described above is subject to some errors, since the time needed between successive input pulses to the counter would vary with the period of the mains supply rather than remaining constant. To avoid this error, at some hardware cost, a voltage-controlled oscillator generating about 25.6 kHz can be used, followed by a nine-stage binary counter. The counter output should be 50 Hz, and this can be locked by a phase lock loop so that the zero of the counter is coincident with the positive-going zero crossing of the mains supply. Pulses from the oscillator will then be controlled so that exactly 256 cycles occur in one half-cycle of the mains waveform. By connecting the counter output and the digit input (assumed to be 8 bits) to an 8-bit comparator, an output pulse can be obtained when the two numbers are equal. This can be used to trigger a thyristor used as a controlled rectifier. This arrangement ensures that if we request a trigger signal after a delay of say 50/256 of a half-cycle, using a digital input of 50, we shall have the same proportional delay and the same mean output even though the mains frequency is not exactly 50 Hz.

A major consideration when connecting a microprocessor system to devices such as thyristors is the need for circuit isolation. Thyristors are often connected between the mains supply and the load and consequently one side of the trigger signal may be at mains potential. To provide adequate isolation from the mains supply either pulse transformers or optical isolators are generally used. These provide insulation rated at 2–3 kV between the mains and the trigger circuits, and have very low capacitance. Thus with a transistor or two to provide current gain, there is sufficient attenuation between the mains and the output package to prevent mains transients from reaching the microprocessor.

Thyristors are normally used as controlled rectifiers to provide variable d.c. supplies; where controlled a.c. supplies are needed triacs can be used. These act like two thyristors back-to-back and so can be used to delay the conduction of both half-cycles of the supply. Thus trigger pulses are needed in every half-cycle of the mains waveform rather than every other half-cycle as for thyristors.

In addition to the control of the triac output that can be obtained by delaying each trigger pulse after the zero crossing, another procedure can be used which is suitable for loads that have a large inertia. This is called 'burst firing' and can be used, for example, in the control of large heating elements.

The control arrangements for this scheme are simpler than those used with variable delay triggering, since the firing pulse is always generated at the beginning of each half-cycle. Control of the power supplied to the heater is obtained by interrupting the supply for a whole number of cycles. Thus if 6 trigger pulses were emitted, followed by 14 half-cycles without pulses, power would flow for 3 cycles out of every 10. This would give a duty cycle of 30 per cent, and the mean power averaged over 10 cycles would be 30 per cent of the power available with continuous conduction.

This procedure cannot give such precise power control as can delayed trigger operation, but is simpler to implement. Owing to the relatively long period of the power control cycle, in this case 1/5 s with 50 Hz mains, the scheme is only practicable if the inertia of the load is large enough to smooth out fluctuations in the power output. The major application of burst firing control is to heating systems where the thermal time constant is vastly greater than 1/5 s, and the temperature fluctuations caused by intermittent power input are negligible.

Figures 11.2 and 11.3 show the load current waveforms for delay triggering and burst firing, assuming purely resistive loads. Figure 11.4 gives a typical interface circuit using a pulse transformer for isolation which could be used to trigger either a thyristor or a triac.

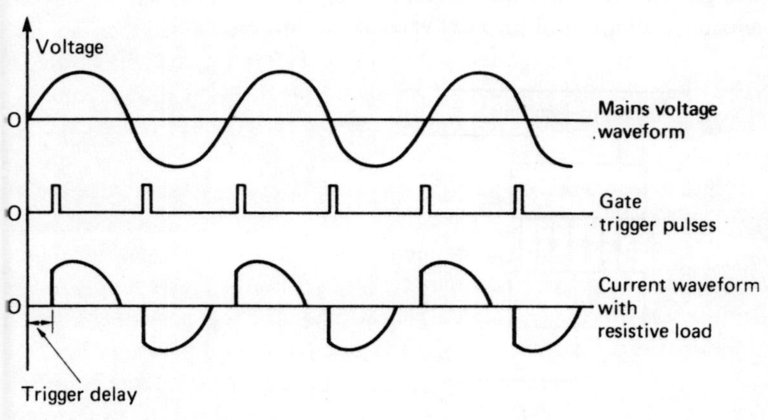

Figure 11.2 Triac control using delayed trigger

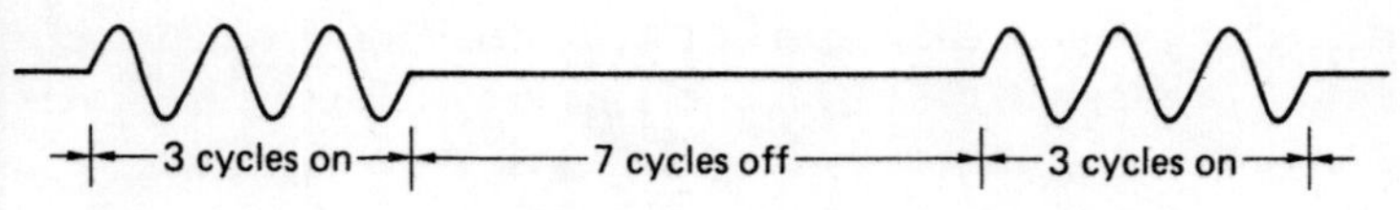

Figure 11.3 Triac load current with burst firing

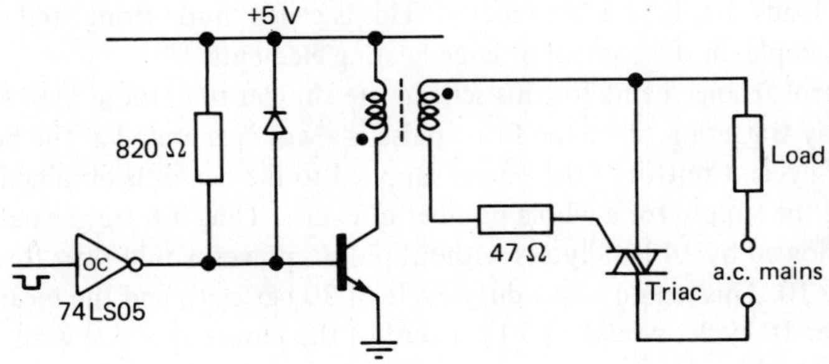

Figure 11.4 Trigger circuit for triac

11.4 Analogue outputs

Analogue output signals are required for driving chart recorders, pointer instruments, oscilloscope displays etc. All of these devices involve only low power outputs which can generally be provided from an operational amplifier connected to the output of a DAC. Where some voltage gain is required without a polarity inversion, the circuit shown in figure 11.5 can be used. An 8-bit DAC can be connected directly to one port of a parallel I/O package; 10 or 12 bit DACs will require 2 or 4 bits of a second port. Since the port outputs are latched, any value sent to the port will be converted into the corresponding output voltage by the DAC and this will remain constant until the next value is output.

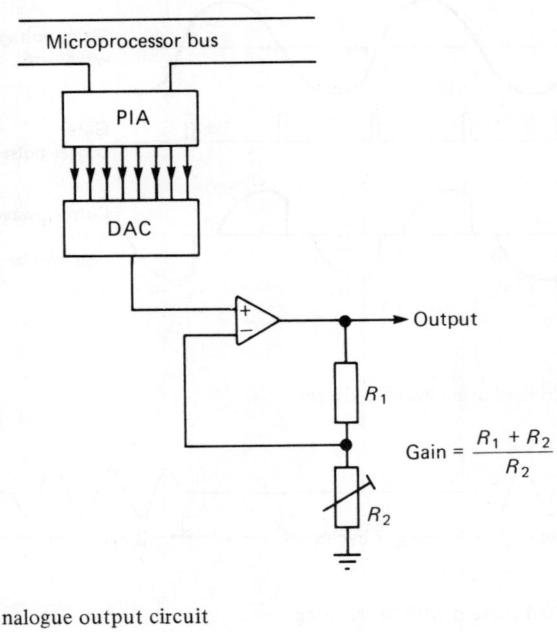

Figure 11.5 Analogue output circuit

If 8-bit resolution is adequate, it may be possible to avoid using a parallel I/O package. Some 8-bit DACs have registers on the digital inputs which allow any value input to be retained. These can be connected directly to the microprocessor data bus. The strobe that latches data into the DAC can be derived from some address decoding together with appropriate timing signals obtained from the microprocessor bus.

For example, if no more than 15 I/O device addresses were required in a Z-80 system, it would be necessary to decode only the lower four address lines, together with $\overline{\text{IORQ}}$ and $\overline{\text{WR}}$. This would generate a pulse for latching data from the microprocessor bus into the DAC register. An 8-way NAND gate and a hex inverter can perform this function and will cost much less than a PIO parallel I/O package.

Most oscilloscopes can handle much higher data rates than microprocessors can deliver, but the reverse is true for chart recorders and pointer instruments such as voltmeters. With these devices the maximum frequency that can be resolved is of the order of a few Hz, so there is no point in providing high data rates. Accordingly it is possible to dispense with a DAC and generate by program a pulse width modulated output. If the repetition rate of this is around 50 Hz, the inertia of the instrument, perhaps with a little electrical smoothing, will serve to eliminate fluctuations in the deflection. This procedure saves the cost of a DAC but requires more use of the processor.

11.5 Stepping motors

Conventional d.c. motors have a major disadvantage when used in position control systems; there is no accurate relation between the voltage and current supplied to the motor and the resulting angular movement. Consequently some means of measuring the position of the load must be provided and this must continuously be compared with the demanded position. The difference between the two can then be used to control the motor power in such a way as to reduce the error.

The system now forms a closed loop feedback control with attendant problems of stability and accuracy. A much simpler system could be devised if there were a precise relationship between the power input and the motor rotation, since this would avoid the need for a position measuring system.

The stepping motor is such a device which is available for low power applications. It comprises a toothed permanent magnet rotor and a wound stator which usually contains two centre-tapped windings. By energising the windings selectively, the rotor can be moved in exact angular increments, for example, of $7\frac{1}{2}°$ or $15°$. By coupling this to suitable gearing, any required step of angular rotation can be provided. Also by a combination of gearing and a toothed rack, a quantised linear motion can be obtained. Any size of step can be provided by a suitable choice of gearing and rack pitch. The only requirement, apart from power amplifiers, to drive the motor windings is a reference position sensor. This can be either an optical device or a micro-switch.

When the system is switched on the motor is moved to the reference position, and is stopped there by the position sensor. A position counter is then cleared, and thereafter the microprocessor program will adjust the counter contents each time the motor is stepped to a new position. The motor can move in either direction depending upon the sequence of pulses. If the application permits, the motor can be returned to the reference position from time to time, to check the accuracy of its position counting.

The microprocessor program must have a degree of complexity since, if the motor is to follow the sequence of driving pulses accurately, it can accelerate only at a limited rate. Thus as soon as the program is aware of the distance to be moved, in terms of the number of steps, it will divide this into three stages – acceleration, constant speed motion and deceleration – and will have fixed maximum values stored for all three. This ensures that the movement will take the minimum of time, consistent with accurate stepping.

A major limitation to the maximum speed is the inductance of the windings which prevents a rapid rise and fall of the driving current.

In order to allow more rapid operation it is customary to overdrive the coils. This means applying initially several times more voltage than the rated value. Some current sensing is included, and when the current has reached the rated value the voltage applied is reduced to hold the current constant. This avoids overheating the stator windings but gives a useful increase in maximum stepping rate. A similar result can be obtained by feeding the coils via a *CR* network.

Driving currents of several ampères at 20–60 V are typical for small motors. This can be provided by Darlington power transistors or by power field effect transistors. The latter have the advantage of requiring negligible input current and can normally be driven directly from an I/O package. This requires four output lines, one for each phase, and with the connections shown in figure 11.6 the current in each winding is reversed in turn in order to step the motor round.

The current reversal is obtained by energising only one-half of each centre-tapped coil at a time. Thus if initially current is flowing in coils 1 and 3, the first step is to switch the current from coil 1 to coil 2, then from coil 3 to coil 4, etc. If the current is turned on by using buffers such as the 4050 which do not invert the signal, an input of logic 1 will turn on the current. The signal sequence required is then as shown in the table below

Coil no.	1	2	3	4	*Forward*	*Reverse*
Driver	1	0	1	0	↓	↑
Inputs	0	1	1	0		
	0	1	0	1		
	1	0	0	1		
	1	0	1	0		

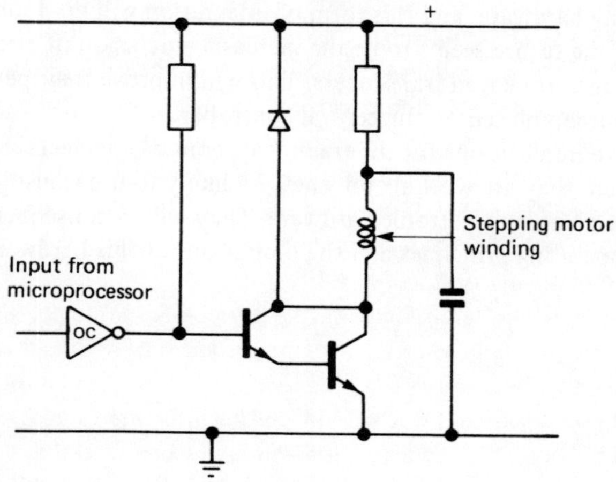

Figure 11.6 Drive circuit for one phase of stepping motor

If this sequence of currents drives the motor forward, the reverse sequence will drive the motor backwards.

The 4-bit patterns required can be stored within the program and called up in turn.

An alternative method of energising the motor coils is to use special-purpose integrated circuits. These include pulse sequence circuits, current control and two driver stages. The only external signals required are a pulse train to control the stepping rate and a direction signal to control the direction of motion. In addition, two other signals can be used to control the current level at which the drivers limit. These allow for current levels to be specified, generally 0, 1/3, 2/3 and full-rated current.

11.6 Future prospects

Although some of the transducers described in this book, and their operating principles, have been known for many years, the technology of transducers is a developing one and new techniques are continually evolving. As control systems are incorporated into a wider range of products and processes, so the variety of transducers attached to them must correspondingly be extended.

A particular trend is for more computation and signal transformation to occur at the transducer and less at the central controller. This suggests a future in which a microprocessor is incorporated into each transducer.

We are at present a long way from this stage, but there are already examples of it in many complex measurement systems. Transducers, particularly of pressure and force, are being fabricated on silicon chips together with amplifiers and other signal conditioning hardware, and this form of integration will no doubt continue. The prospect for the future seems to be the increasing inclusion of electronic hardware of all kinds into packaged transducers. This will improve their performance and reduce the burden placed on the central controller.

It will also have implications for the training of control engineers since, whatever type of system they are working on, they are likely to use transducers that incorporate quite complex electronic hardware. They will in consequence need to understand the operating principles and the limitations of this hardware.

Bibliography

Adams, L. F., *Engineering Measurements and Instrumentation*, E. U. Press, London, 1975

Auslander, D. M. and Sagues, P., *Microprocessors for Measurement and Control*, Osbourne/McGraw-Hill, Berkeley, California

Benedict, R. P., *Fundamentals of Temperature, Pressure and Flow Measurement*, Wiley, London, 1969

Bibbero, R. J., *Microprocessors in Industrial Control*, Prentice-Hall International, Hemel Hempstead, 1984

Carr, J. J., *Elements of Electronic Instrumentation and Measurement*, Reston Publishing Co., Reston, Virginia, 1979

Clayton, G. B., *Data Converters*, Macmillan, London, 1982

Cluley, J. C., *Interfacing to Microprocessors*, Macmillan, London, 1983

Collett, C. V. and Hope, A. D., *Engineering Measurement*, Pitman, London, 1974

De Sa, A., *Principles of Electronic Instrumentation*, Edward Arnold, London, 1981

Earley, B., *Practical Instrumentation Handbook*, Scientific Era Publishers, Stamford, Connecticut, 1976

Gregory, B. A., *An Introduction to Electrical Instrumentation and Measurement Systems*, second edition, Macmillan, London, 1981

Kochhar, A. K. and Burns, N. D., *Microprocessors and their Manufacturing Applications*, McGraw-Hill, New York, 1983

Lenk, J. D., *Handbook of Microcomputer Based Instrumentation and Controls*, Prentice-Hall International, Hemel Hempstead, 1984

Mansfield, P. H., *Electrical Transducers for Industrial Measurement*, Butterworths, London, 1973

Morris, N. M., *Microprocessor and Microcomputer Technology*, Macmillan, London, 1981

Neubert, H. K. P., *Instrument Transducers*, Oxford University Press, Oxford, 1975

Norton, H. N., *Handbook of Transducers for Electronic Measuring Systems*, Prentice-Hall International, Hemel Hempstead, 1969

Oliver, F. J., *Practical Instrumentation Transducers*, Pitman, London, 1972

Sarshe, R. B., *Semiconductor Temperature Sensors and their Applications*, Wiley Interscience, London, 1975

Seippel, R. G., *Transducers, Sensors and Detectors*, Prentice-Hall International, Hemel Hempstead, 1983

Sonde, B. S., *Transducers and Display Systems*, Tata McGraw-Hill, New Delhi, 1977

Spitze, F. and Howard, B., *Principles of Modern Instrumentation*, Holt Rinehart and Winston, New York, 1972

Sydenham, P. H., *Transducers in Measurement and Control*, University of New England Publishing Unit, Armadale, Australia, 1975

Tomkins, W. J. and Websters, J. G. (Eds), *Design of Microcomputer Based Medical Instrumentation*, Prentice-Hall International, Hemel Hempstead, 1981

Walker, B. S., *Understanding Microprocessors*, Macmillan, London, 1982

Wightman, E. J., *Instrumentation in Process Control*, Butterworths, London, 1972

Woolvet, G. A., *Transducers in Digital Systems*, Peter Peregrinus, London, 1977

Index